MOSTINI Planet Next Level

The World with Josammy Technology
– Alpha & Omega Thermodynamic
Sigma ZG Matrix

Volume No. 1

Josammy Samba Ganga – Josammy
Emporio 3D

MOSTINI Planet Next Level

Copyright © 2020

All Rights Reserved

MOSTINI Planet Next Level

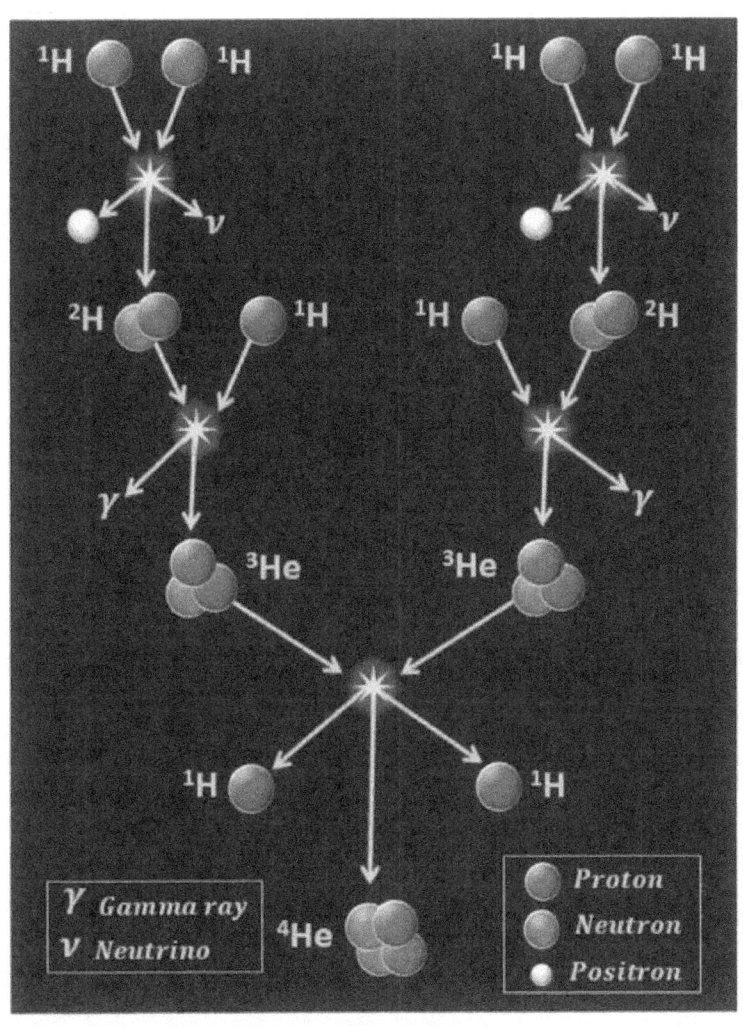

MOSTINI Planet Next Level

Dedication

It is with the deepest gratitude and warmest affection that I dedicate this book to scientist, philosophers, and laymen. I dedicate this book to lovers of sciences, knowledge, and universe. I hope this book helps you to achieve your goal, understand all universal reactivities and creation, solving all mysteries of this universe. This book helps you to get work out to a wider audience, which it allows to stir hearts and mind of many people.

Reading this book will allow you to dive into a universe where the understanding of many scientific fact of our universe will no longer hold any secrets for you. You will thus answer countless questions that you faced throughout your life without any concrete explanation. After reading this book, you will see this world from another angle. Nothing will be the same.

I would love nothing more than to see this book in hands of world people, learning in Mostini planets school.

MOSTINI Planet Next Level

"You will know the truth and the truth will set you free"

I am very grateful to my parents and the people who helped me to accomplish this wonderful work (Vox Ghostwriting)

MOSTINI Planet Next Level

Acknowledgment

This book is the greatest wonder that is accomplished to rehabilitate and satisfy human curiosity in all levels of their life. It is notably meant for laymen, scientists, and philosophers who are focusing on the hardest and critical questions relating to our existence and our presence on the planet.

The book is about to explain real universe reactivities. I am illuminating this world on the Gravity knowledge, since the beginning scientist , philosophers, laymen never understand correctly the Gravity, all fields of science are misinterpreted . This book solves all scientific enigmas all mysteries of universes , displays and satisfies curiosity of scientists, philosophers and laymen on Gravity knowledge. The book is going to explains reactivities of formations of planets, Solar systems, Galaxies and whole universe with impact influence of highest cosmic laws " Alpha & Omega Thermodynamic Matrix Sigma ZG and the Quantum Matrix Antigravity Sigma ZG

MOSTINI Planet Next Level

Since the beginning of time, our ancestors never felt happy when considering all the unanswered questions there were concerning their existence on Earth. Since primitive times, in the great struggle that enabled man to free himself from the animal kingdom, Nature was the first shepherd of humanity and its first enemy.

For billions of years, Nature has always oppressed and subjected men to its laws, such as the fundamental, thermodynamic, centrifugal force and the centripetal force's microcrack extension stress stability.

To free himself from this natural oppression, man has perpetually devoted himself to the conquest of the world. His essential tools for this task were found in the form of philosophy, science and technology.

Slowly, a new world arrived quickly – one that was more scientific and effective. There are more treatments available for old diseases. We believe in having a better future that is powered by the people who have the courage to innovate this world with their creativities.

People like *Boltzmann, Ricci, Euler, Nernst,*

MOSTINI Planet Next Level

Newton, Helmholtz, Auguste Comte, Kepler, Nicolas Tesla, Einstein, Avocado, Coulomb, Van Der Waals, and now, Josammy Emporio 3D. Our natural love for knowledge engenders us towards technological advances that are filled with light. It is as natural as the desire to sing.

Everything around us moves with its own energy.

The sea that reveals itself to show it murmurs is the need to love. All these Stars that fall right, it is the chance of you, of you who will come to make shine my City and to hear it singing. And the horizon that lights up in the corner of your eyes is a lot of desire.

The Cloud that files beyond the night is the dream that dreams. And this sky is too big to hide our secrets, it is the immense of you, of you who will come to lose in my city and the hollow of my arms.

Remember to look up at the Stars and not down at your feet. Try to make sense of what you see and wonder about what makes the Universe exist. Be Curious

O pain! O pain! Time eats life, and the dark enemy that gnaws at our heart. Blood that we lose is growing and

MOSTINI Planet Next Level

strengthening.

Nature teaches us nothing useless. It is also Nature which compels man to kill, to eat, to sequester, and even to torture his fellow man. As soon as we leave the natural order of necessities and needs to enter into the fields of luxury and pleasure, we see that Nature can lead one to crime.

Nature is nothing but the voice of our interest. Virtue, knowledge, and philosophy, on the contrary, are artificial, almost supernatural. Evil is effortless.

Of course, good is always the product of art. The virtues of this book is that it welcomes you to a *Mostini Planet School* that encourages a technology and information transfer.

"you will know the truth, and the truth will set you free."

MOSTINI Planet Next Level

Contents

Dedication 4

Acknowledgment 6

About the Author 24

Preface 26

The Development Principle Of Knowledge 43

Technology Transfer & Consciousness Development 47

Key Concepts of the Reality Approach of the Universe 54

Lost in This World 56

Chapter I 60

Mostini Planet 60

Question 1: What Is The Universe Made Of Or What Is The Composition Of Our Actual Universe? 62

Question 2: What are The Dimensions of the Universe and the Structures Parameters? 62

Question 3: What is the Expansion Of the Universe

MOSTINI Planet Next Level

And Why Is It so Easy Destroy?..63

Question 4: Why Does Life Exist? 64

Question 5: Can Matter Escape Black Holes? 65

Question 6: Why Are There Very Cold Zones and Very Hot Zones and Why Is the Universe Adiabatic? 66

Question 7: Do Entangled Particles Imprint the Universe Laws? 67

Question 8: What is Time? 67

Question 9: Are We Alone in the Universe? 67

Question 10 - Do We Have A Distorted View Of The Universe? 68

Question 11: What Are the Highest and the Greatest Universe Forces? 68

Question 12: What are the Highest Laws Reactivities of the Universe? 69

Space Travel Adventure Inspection 70

Alpha & Omega Thermodynamic Sigma ZG Matrix and Thermodynamic Function 73

MOSTINI Planet Next Level

Impact and Influence of Systems and 2nd Thermodynamic Laws in The Living System and Universal Entities...75

Electricity Origin, Activities And Reactivities In Living Organisms (Humans, Animals, Plants)...................80

Alpha & Omega Thermodynamic Sigma ZG....................82

Chapter II 86

The Quantum Matrix Antigravity Sigma ZG Potential Sigma ZG..86

The Potential Sigma Z 87

The Potential Energy ZG 91

The Physiochemical Potential Sigma ZG 93

The Geo-thermodynamic Sigma ZG 95

Chapter III Josammy Technology Application..................97

The Thermodynamic Reactivities of Hydrogen 97

The Fundamental Thermodynamic Force Properties, Activities Law Reactivities And Description 99

The Reactivities Description Of the Universe

MOSTINI Planet Next Level

Primordial Or Primitive 101

The Zeta Zero or Point Energy 102

The Universal Black Energy 103

Chapter IV Barrier Energy and Thermodynamic Regulation 106

1. The Thermodynamic Matrix Sigma ZG Stability and Barrier Energy 106

2. Zeta non-Zero and The Quantum Matrix Antigravity . 108

3. The Fundamental Thermodynamic Force Regulation in the Universe (Microcrack extension stress stability) 110

4. Thermodynamic Regularization by Neutralization 112

5. Regulation by the Principle of Action and Reaction ... 117

MOSTINI Planet Next Level

6. Electronic Regulation Sigma ZG................120

6.a. The Fundamental Thermodynamic Force and the Particle In Electric Field 120

6.b. Geographical Area Regulation or Geo-thermodynamic 125

The Quantum Metrix Antigravity 130

7. The Potential Physicochemical Sigma ZG Regularization and Physical Impact 130

8. Thermodynamic Regulation by Friction Sigma ZG 135

9. The Fundamental Thermodynamic Force Regulation sigma Z G by Influence or levitation 137

10. Geo-Thermodynamic Sigma ZG Riemanndeformation Regulation 139

11. The Mechanical Sigma ZG Thermodynamic 144

Regulation 144

The fundamental thermodynamic force mechanic Sigma ZG Regulation and projectile 144

MOSTINI Planet Next Level

12. The fundamental Thermodynamic Sigma ZG Heavy Mechanical Pendulum and Oscillation Regulation..................................146

13. Osmosis and Thermodynamic Regulation.................150

Interpretation Thermodynamics..155

14. Coulomb's Work by the Principle Of Regulation......156

15. The Fundamental Thermodynamic Force Regulation and Helium Balloon 159

16. The Fundamental Thermodynamic Force And Buoyancy 161

17. The Thermodynamic Properties and Regulation Between Earth and Moon 163

18. The Astronaut Thermodynamic Properties In The Space 168

19. The Properties Of The Fundamental Thermodynamic Force Of The Apple On The Moon And On The Earth..173

MOSTINI Planet Next Level

20. The Thermodynamic properties between earth and moon 180

21. Manifestation and Influence of Second Thermodynamic in The Living and Human Body (The second Thermodynamic Law) 181

22. The properties of the fundamental thermodynamic force and IMC or BMI body mass index(2nd thermodynamic law regulation).......................................186

23. Thermodynamic Law Mineral Regulation Mineral Through Organic (Ortho-silicic acid In The Formation of Living Cell, DNA, and Life the Genetic Takeover)....191

24. Electricity regulation activities in the living system(Animals and human) 194

25. Thermodynamic gas parameters regulation...............196

26. The physicochemical potential Sigma ZG Energy Barrier Zeta...199

Chapter V The Pollution of the Universe 201

1. Regulation of Universe Pollution Charge....................201

2. Geo-Thermodynamic Lightning Phenomena on the

MOSTINI Planet Next Level

Effects of Annihilation and Regulation 202

3. Thermodynamic universe regulation 205

4. Geo-thermodynamic or Potential Sigma ZG Regulation and Kinetic Moment......................................205

5. Potential Sigma ZG and Magnetism Regulation 207

6. Potential Sigma and Radioactivity Regulation 209

7. Potential Sigma ZG Regulation and the Reversing Magnetic Earth...210

Chapter VI   The Fundamental Thermodynamic Force And Universe Entropy Reactivities And Functioning.....217

Entropy Reactivities..217

1 Mechanical Entropy..220

2 Kinetic Entropy...220

3 Potential Entropy 222

The Potential Thermodynamic 225

5 Geothermal of Earth Planets (The Heat and Entropy of Earth planets)...227

MOSTINI Planet Next Level

6 The Potential Thermodynamic Force 239

The Rotation Centrifugal of Riemann 240

Chapter VII Thermodynamic Scale 243

1 Universal Thermodynamic Scale 243

2 Universe Description 243

3 Universe Thermodynamic Scale 245

3a Interpretation 246

3b Reactivities Thermodynamic scale of Moon, Earth planet, Jupiter, and Sun..................................247

3c Interpretation 248

4 Analyzing Thermodynamic Reactivities Of Our Solar System 249

4a Analyzing Thermodynamic Distance Between Our Sun And Jupiter 249

4b Analyzing Thermodynamic Reactivities between 251

4c Analyzing reactivities between Jupiter and Neptune..253

5 Quantum Entities Identification (QEI) 255

MOSTINI Planet Next Level

5a Security Identification 256

5b Identification of Entities by the QEI method..............257

6 The Folding Universe 259

7 Mars Planets Irrelevance Thermodynamic Status 260

7a Interpretation..261

8 The Mathematical Language Of Universal Reactivities 263

8a Structural...264

9 The Universe Primordial Thermodynamic Reactivities 269

10 The Geo-thermodynamic Form Reactivities 273

11 The Universal Geo-Thermodynamic Scale Reactivities 276

12 Oxidoreduction Reactivity Through Universe Reactivities 277

13 Comparison of the Moon (Earth Satellite), the Planet Earth, Jupiter, Saturn and Stars (our Sun) 280

14 Geo-thermodynamic rotation charge identification....282

MOSTINI Planet Next Level

14 a. Quantitative 284

15 Quantum universal Entities Reactivities (QUER) 287

15a The rotation of Celestial entities around Themselves 287

16 The universal force reactivities scale.........................290

17 Interpretation..291

18 Why All The Planets Of Our Solar System Turn in the Same Direction Except Venus, Uranus, and Neptune...294

MOSTINI Planet Next Level

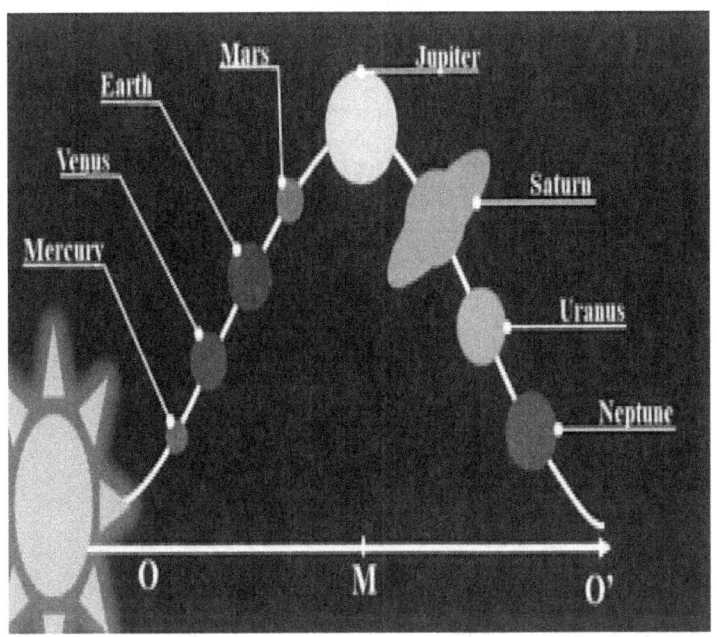

MOSTINI Planet Next Level

About the Author

Josammy Ganga Samba is an American researcher who lives in San-Diego, California, United States. Born April 13, 1969, from an early age, he is passionate about discovering and understanding the facts around us. It is with this in mind that he travels through different establishments and universities to achieve his goals.

Today he is offering us a book which will be subdivided into three (3) volumes and which will be titled: Mostini Planet, Next Level the World with Josammy Technology (The quantum Antigravity Sigma ZG Matrix).

He has five patents to his name, four of which are in the US and one is international. Josammy Samba Ganga is the first man in the world history to discover the real reactivities of the highest cosmic laws. Cosmic laws that govern the universe.

MOSTINI Planet Next Level

Preface

Stellar formation is an area that arouses interest not only because of the complex and mysterious phenomena which take place there, like the accretion and the ejection of the materials or emission of rays but also because it is related to the questions of our origins.

Understanding the genesis of stars and their planetary system tells us about the history of the solar system and about the primitive chemistry that occurred during the formation of the Earth and the appearance of life.

Universe space is the result of ignorance of the unseen energies that occupy the space around us. Obviously, certain light frequencies, or the tiniest particles of matter that makeup atoms are invisible to us. Energy is often invisible, such as electricity or movement of the air. We only know about its presence when it interacts with a denser, physical object.

Everything in the Universe is related through energy, seen and unseen. We were born with the ability to feel this energy. Our bodies are precise, accurate, and sensitive instruments that measure and communicate this energy with everyone around us. Also, we live in this Universe because everything is just a living system made with energy, including the animals and plants.

MOSTINI Planet Next Level

It is abundant energy that would come from an infinite source that does need to be centralized before being redistributed by various means. The Universe is just an inverted and spherical plasma screen that functions with energy powered and governed by the fundamental thermodynamic centrifugal force reactivities and microcrack extension centripetal force.

Everything in this Universe is just energy. It appears that the Creator of this Universe has a strong point to make regarding that – the standing waves represent the electric potential that covers every speck of space. In other words, the Creator's energy is ever-present throughout creation. And this presence is always there to assist us, but we must learn how to tune in to it properly.

There are scientific methods of meditation that can help take those, who are ready, to the next level with tools like the *Mostini Planet School* and the *Josammy Technology*. Obviously, the Sun and our solar system orbit the center of the Milky Way galaxy.

We are always moving at an average velocity of 828,000km/hour. But even at that high rate, it still takes us about 230 million years to make one complete orbit around the Milky Way!

The Milky Way is a spiral galaxy due to our human distorted vision. We believe that it consists of a central bulge, four major arms, and several shorter arm segments.

MOSTINI Planet Next Level

The sun and, of course, the rest of our Solar system, is located near Orion's arm, between two other majors arms namely, Perseus and Sagittarius. The diameter of the Milky Way is about 100,000 light-years, and our Sun is located about 28,000 light-years from the Galactic center.

- The Gravitational centripetal force attraction is proportional to nuclear energy stored in the mass entities
- The Gravitational centripetal force attraction level exerts between two or several universal entities are inversely proportional to each other , and relates their distance mass placement in equilibrium to lowest energies in search of zeta energy=0
- The fundamental thermodynamic centrifugal level exerts between two or several universal entities are inversely proportional to each other , and relates their mass placement in equilibrium to lowest energy in search of zeta energy=0

Obviously, our planet, Earth, rotates and revolves at 30 km per second. The Earth rotates once every 23 hours, 56 minutes, and 409053 seconds. It is also called the sidereal period.

Its circumference is roughly 40,075 km, and at the

MOSTINI Planet Next Level

equator, the Earth moves at a speed of 460 meters per second or roughly at 1,000 miles per hour. Despite this, it takes the Sun one year, or 365 and 1/4 days, to orbit the Sun completely.

As the Earth orbits the Sun, the Moon orbits the Earth. In the Universe, and in our Galaxy, some planets revolve around two, three, and even four stars. These are known as *Multisystem Stars*.

All eight planets in our solar system orbit around the Sun and are influenced by its rotation, which is counterclockwise. When viewed from above the Sun's north pole. Six of the planets also rotate about their axis in this Same direction. The exceptions here are the planets with retrograde rotation which are Venus and Uranus.

Venus's axial tilt is at 177 degrees, so its axis of rotation is approximately parallel with the plane of the Solar System. Planets in multi star systems *(Mostini Planet)* have special interest to mankind, including astronomers and planetary scientists, because they provide an example of how planet formation functions, and really explain the reactivities of the Universe by intervening and implying that the principle process reactivities of the potential physicochemical *Sigma ZG*, the Geo-thermodynamic *Sigma ZG, Alpha & Omega Matrix Sigma ZG*, the *Quantum Matrix Antigravity Sigma ZG* also explain the reactivities and the reason of Uranus's unusual

MOSTINI Planet Next Level

axial spin.

Multi-star systems seem to be exotic to us in our orbit around our solitary star. Multi-star systems imply the connection of the four interconnected *Sigma ZG Josammy's technologies*.

Newton's law of universal gravitation is stated as, *"Every particle attracts other particles in the Universe using force that is directly proportional to the product of their masses and inversely proportional to the square of the distance between their centers."*

Obviously, Newton realized that the force of gravity here on Earth that causes things, likes apples, to fall works in the same way everywhere else in the Universe too. His work proved that gravity worked on celestial objects, like the Moon, the Earth, and the planets, and was able to calculate their orbits, which had long been observed.

Einstein offered a different view of gravity, one that made sense of Mercury's movements. Instead of exerting an attractive force, he reasoned that each object curves the fabric of space and time around it, forming a sort of well that other objects, and even beams of light, fall into. Actually, scientists claim that Newton's law of gravity is inconsistent, so they target Einstein's theory of general relativity next.

The world learned about the famous law of gravity when an apple fell on Isaac Newton's head, prompting him

MOSTINI Planet Next Level

to form the earliest theory of universal gravitation. However, most researchers think that general relativity is wrong. To be precise; most believe it is *incomplete*. After all, the other forces of Nature are governed by quantum physics; gravity alone has stubbornly resisted a quantum description.

The fix is dark matter, particles invisible to light but endowed with gravity. Considering Newton's theory and Einstein's theory, many corrections should be made by intervening and implementing the highest cosmic law, such as the fundamental thermodynamic centrifugal force that governs the Universe. The fundamental thermodynamic centrifugal force and microcrack extension stability centripetal *(Entropy)* are inversely proportional, to be more precise, and solve many enigmas.

Many mysteries that escape the comprehension of the functioning of the Universe can be solved with their help. Obviously, we still have a lot of puzzles and contradictions in the comprehension of the Universe. For instance, our stars (the Sun) is several times larger than all the planets put together.

According to gravity based on Newton's formula and Einstein's theory's reactivities, which implies gravity is a mass of ratio, should be assessed correctly. This is because in the Universe, each entity has its own size, and all entities should collide. So, by taking Newton's gravity

MOSTINI Planet Next Level

into account as a mass of ratio, we are asking why the planets in our solar system do not collide with the Sun.

Also, why, in this same report of the mass of ratio, apple placed on the Moon does not fall? Why apple kept and released at a distance H from the moon's ground does not fall when the same is placed at the same height H on the planet Earth, the apple fall? (The Moon is larger than an apple) In this trend, scientists state that Archimedean law is a force that would be a particular and an exception force, challenging all universal force prescriptions. If gravity is the mass of ratio, the Archimedean theory would not confer an exception.

The understanding of the Universe should imply the comprehension of highest universe laws, namely:

- The fundamental thermodynamic force centrifugal governs Universe.
- The fundamental thermodynamic centrifugal force and microcrack extension centripetal are inversely proportional.

The greatest truth of understanding the Universe in all shapes, dimensions, and reactivities have been proven by the highest universe world theories such as:

- The Alpha & Omega Thermodynamic Sigma ZG
- Quantum Metrix Antigravity Sigma ZG
- The physicochemical potential Sigma ZG

MOSTINI Planet Next Level

- Geo-thermodynamic Sigma ZG

These four disciplines represent the highest technology of the comprehension of reactivities of Universe in all forms and dimensions and allow us to resolve all scientific enigmas and riddles.

We can do this while rehabilitating Newton's and Einstein's theory for ultimate comprehension of the Universe and solving the scientific enigma.

I am the first human in all the world who discovered the ultimate reactivities of the highest cosmic law to know the following:

- The fundamental thermodynamic centrifugal force governs the Universe.
- The fundamental thermodynamic force centrifugal and microcrack extension stress stability (Entropy) centripetal force are inversely proportional.
- The entropy centripetal of system is proportional to the mass entities.
- The mass of system is inversely proportional to the fundamental thermodynamic centrifugal reactivities.
- The Entropy centripetal force and microcrack extension stress stability are proportional.
- The microcrack extension stress stability is proportional to Gravitational centripetal attraction force entities.
- The Gravitational centripetal attraction forces

entities are proportional to mass entities.
- The Gravitational centripetal force entities are proportional to heat energy stored in the mass entities. The Gravitational centripetal forces entities are proportional to light energy stored in the mass entities.
- The Gravitational centripetal force attraction is proportional to radiant energy stored by mass entities.
- The Gravitational centripetal force is proportional to the sound produced by mass entities.
- The Gravitational centripetal force attraction is proportional to potential energy centripetal stored in the mass entities. The Gravitational centripetal force attraction is proportional to Chemical energy of mass entities system.
- The Gravity centripetal attraction force is proportional to electromagnetic force mass entities
- The Gravity centripetal force attraction is proportional to potential energy stored in mass entities
- The Gravity centripetal force attraction is proportional to chemical energy stored in the mass entities
- The Gravitational centripetal force attraction is proportional to electromagnetic force entities stored in the mass system.

MOSTINI Planet Next Level

These scientific realities could be verified by including these law functionalities in all fields of science. Effectively, this innovative theory constitutes base function universe reactivities. Also, solving scientific enigma and satisfy curiosity of mankind in the Universe formation reactivities and functionalities formation.

However, in actual Universe a significant factor that hinder an operate from entering industry is high levels of competition regulation associated with low level technology comprehension of actual universities, companies, and organization because Mankind never understand gravity and the reactivities of highest cosmic laws precisely, the reactivities between microcrack extension stress stability (Entropy), the centripetal force.

Although the fundamental thermodynamic centrifugal governs all Universe in infinitely large as well as in infinitely small, the Big Bang theories and others universal sciences namely: Medicine, Astrophysics, Astronomy, Biology, Chemistry, Geophysics, Geology, Climatology, Seismography, Oceanography...

Seems to be pseudoscience because they don't intervene the implication of highest cosmic laws in their demonstration and theories reactivities. Obviously, with no implication and absence implication of fundamental thermodynamic centrifugal force, no one in the Universe could explain Universe gravity and solve the mysteries of

MOSTINI Planet Next Level

Universe.

Obviously, nobody can explain gravity without intervening the impact of the fundamental thermodynamic centrifugal force because the fundamental thermodynamic centrifugal governs the world with the reactivities inversely proportional with microcrack extension stress centripetal force stability.

The purpose of this book is going is to enlighten this world bringing joy and curiosity satisfaction expected by all people who are living this planet. I discovered reactivities of highest cosmic and Universe law. The fundamental Thermodynamic centrifugal force reactivities and microcrack extension stability centripetal force (Entropy). These highest universal laws are unfortunately known to be misinterpreted of the comprehension formation and the understanding of universe law.

Gravity has been poorly explained since the earliest time. My researches have given opportunities to mankind to understand reactivities of highest cosmic law. To solve all riddles and mysteries creation including life creation, genetic take over or DNA formation. I believe that it would be worth trying to learn something about the universe creation, universe reactivities functioning, highest universe law even if we merely learnt that we do not know much.

This state of learned ignorance might be help in many of our troubles. It might be well to remember that,

MOSTINI Planet Next Level

while differing widely in the variously of little bits we know, in our infinite ignorance we are all equal.

If only we look for it, we can often find a true idea, worthy of being preserved, in philosophical and scientific theories which we must reject as false.

Can we find an idea like this in one of the theories of the ultimate source of the knowledge?

I believe we can; and I suggest that it is one two main ideas which underlie the doctrine that the source of all knowledge is Supernatural. The first of these ideas is false, I believe, while the second is true.

The first, the false idea, is that we must justify our knowledge, or theories by positives reasons, that is the reasons capable of establishing them, or at least of making them highly probable; at any rate, by better reason than they have withstood criticism.

This idea implies as I suggested, that we must appeal to some ultimate or authoritative source of true knowledge; which still leaves open the character of that authority -whether it is human, like observation or reason, or super human *(and super natural)*.

The second idea — whose vital importance has been stressed by Russell-is that no man's authority can stablish the truth by decree; that we should submit to truth is above human authority.

MOSTINI Planet Next Level

Taken together these two idea almost immediately yield the conclusion that the source from which our knowledge derives must be super human; a conclusion which tends to encourage self-righteousness and use force or a radical strategies and proven scientifically to those who refuse to the evidence and truth realities.

The source of our all knowledge is mankind effort, mixed with our errors, our prejudice, our dreams, our hope, our weakness, our sorrow, and expectation. Effectively, humans are naturally intelligent, from early age, all human beings use some mathematics, even in culture that have no written language.

Others animal also use mathematics. Crow have been known to keep track of up to thirty persons. Bees can measure angles and length. And almost all animals learn to recognize shapes and sizes *(yes shapes and sizes are part of math.)* For example, rabbits must learn, the shape of a flying hawk. So that they can take cover. They must also learn the shape of edible leave. For all animals' mathematics means survival.

Obviously, Humans are probably born with some basic mathematical abilities. With no teaching whatsoever, almost anyone can tell the difference on object and two objects, know that one object, know that one object is much larger than another and recognize the difference between circle and triangle. But the higher levels of mathematic

MOSTINI Planet Next Level

requires training.

you must learn special techniques to tell the difference between $F(x)=Log(x)$ and $G(x)= Sin(x)$. Many people, especially those who have *"always have a trouble in mathematics."* I think that is impossible to relax with mathematics. What they probably do not know is that mathematics is a part of life. People usually think of mathematic and evidence as a mental process.

When the object differs in size from the drawing, each line or curve in the drawing must be proportional to the same feature in the object. To have proportion, must keep the same relative size, or scale in all part of your drawing. In the drawing scale, you can work in metric or customary (English) units.

In this trend with desire to understand Universe and its reactivities implies the comprehension of its laws reactivities, which are the fundamental thermodynamic centrifugal force and microcrack extension stress stabilities centripetal, and react with same properties in the microscopic scale as well in macroscopic scale (Infinitely large and infinitely small). If we consider these laws reactivities we are able to compare and determine reactivities by analogy regardless universe size.

Obviously, the universe is so huge and complex, that scientists, philosophers, and laymen could not understand the reactivity of the universe. Throughout

MOSTINI Planet Next Level

history, humans have used a variety of ways to help them answer the question? how far? how big?

The generation of explorers have looked deeper and deeper into the vast expanse of Universe. And the journey continues today. In the third century BC, a Greek philosopher asked a question *"How far the moon is away?"* He was able to measure the distance by looking a shadow of the Earth on the moon during a lunar eclipse.

Obviously, Edmond Halley found a way to measure the distance Sun and the planet Venus. That was 300 years ago (Halley is a famous for predicting the return of comet that is named him.) He knew that the planet Venus planet would pass directly between Earth and the Sun usually every 121 years. Did you catch the last time this happened? It was June 8, 2004 and if you missed it in 2012. After that, there will not be another Venus transit until 2117.

Beyond our Galaxy lie many more galaxies. The most distant of galaxies are so far way, the light for them that is hits Earth today set out from galaxies billions of years ago. So, we see them not are today, but as they looked long, they were any life on the Earth. The Universe could be infinite, both in term of space and time, but there is currently no way to test whether it goes on forever or is just big.

The part of the Universe that we can observe, is finite, coming to about 49 billion years. We know we have

MOSTINI Planet Next Level

new methods. And we are making new discoveries. The ultimate methods are designed to comprehend the highest cosmic laws which give the option of interconnection reactivities between infinitely small and infinitely and inter-connection between the microscopic scale and macroscopic scale.

The observation of big and small in the Universe implies connection analogies between mathematical and other branches of science. The notions of *"big"* and *"small"* in our science are related to numerical methods of recognizing universal things around us. It is one of the mathematical, cognizing ways, so we have very big such as *1000M 000* and extremely small such as *1/1000A 1000*- *"extremely big"* and *"extremely small"* in mathematics.

When naturally and logically have infinite " infinite big (infinity)" and infinite small (infinitesimal)." In fact, I undertook research which resulted in me obtaining five patent applications (U.S. patents and international PCT) with more than 70 new biochemical formulation reactivities, and finally the objective comprehension of reactivities of the fundamental universe laws reactivities, and the genesis reactions of all the universal entities including genetic takeover

MOSTINI Planet Next Level

The Development Principle Of Knowledge

The main aim of scientific research is to examine the strategies that can be matched to increase the effectiveness of the knowledge in cosmology, all fields of life, and the development cycle in manufacturing and operations works.

Obviously, modern cosmology is valuable for what it reveals about the nature of the cosmos we inhabit. It is the demonstration of the power of modern science to transform our understanding of who we are and where we came from.

However, most cosmologists focus largely on scientific questions and are not fully aware of the impact of cosmological theories on culture, including politics and arts.

This book introduces this wider context because both scientists and the general public should be aware of the broader importance of their work and its influence on the way we think.

Cosmologists often rely on their fascination with the subject. In the fields of astronomy and cosmology, we live in a period of excitement. Cosmology and astronomy impact culture and is described and represented by it.

If we select four fundamental causes of changes in

MOSTINI Planet Next Level

our perception of the world in the last century, then they would be first relativity, second quantum mechanics, third expending of Universe and fourth, the space program.

The first free date from a fairly narrow time band, if we date them, could be as follows:

- Special relativity from 1905
- General relativity 1915,
- The belief that the Universe is expanding and is much bigger than previously believed through Edwin Hubble's publications from 1924 to 1930
- The quantum mechanics from Niels Bohr and Werner Heisenberg's formulation of the Copenhagen interpretation 1925-1927.

This epic revision of scientific knowledge of underlying structures of the Universe was therefore concentrated into just a quarter century. The dramatic period of the human space program was concentrated into just over 8 years spanning from the first human space flight in 1961 to the Moon Landing in 1969.

All these fundamentally altered the way that we think about life here on Earth. Often these changes are taken for granted. For instance, mobile phone technology, dependent as it is, on satellite networks, is transforming not only the social lives of teenagers in the world but also, the economic muscle of all social economy.

MOSTINI Planet Next Level

Meanwhile, super-fast quantum computing uses phenomena such as entanglement driving and is driving the development of superintelligence, and hence of robotics. The implications of change for society over the next decades are potentially enormous.

The most important conclusion to be drawn from this combination of revolutionary changes is from the role of an observer, as the basis for differing perspectives of time and space relatively, an influence on the world in quantum mechanics, and as a witness for the first time, of the spherical Earth, hanging in space, in a photograph taken by Apollo astronauts in 1968.

Such ideas and experiences have decisively underpinned modern ideas that one person's complete individual experience or perception is as equally valid as anyone else's. Einstein is held particularly responsible for these ideas as a result of his popular equation between relativity on the hand and cultural relativism *(the idea that no one culture is superior or inferior to another)* on the other.

Moral relativism *(the idea that no one culture is morally superior or inferior to another)* is controversial and widely rejected, but cultural relativism does have beneficial scholarly consequences.

This is especially the case in the new field of cultural anthropology. Researchers are required to abandon

MOSTINI Planet Next Level

the idea that one culture is superior or inferior to another in order to better understand other cultures.

Although cosmology and astronomy have improved since the beginning of time, successful negotiation of everyday life would seem to require that people and scientists should possess insight about the deficiencies in their intellectual skills.

However, people tend to be blissfully unaware of their incompetence. This lack of awareness arises because poor performance is doubly cursed: Their lack of skills deprives them not only of the ability to produce correct responses, but also of the expertise necessary to understand that they not producing them.

People base their perceptions of performance, in part, on their preconceived notions about their skills. Because these notions often do not correlate with objective performance, they can lead people to make judgments about their performance that have little to do with actual accomplishment.

Obviously, in the scientific skill development product, the effectiveness of different organizing strategies to enhance the quality of the manufacturing process and product is well established.

In knowledge works, we lack such a framework. Unlike in manufacturing and operational processes, the knowledge development process is often chaotic, unstructured, and unsystematic, resulting in intangible products.

Therefore, the principle of manufacturing strategies should be defined and initiated based on the knowledge development phase (e.g., knowledge creation, knowledge adoption, knowledge distribution, and knowledge review and revision).

Each phase in the development cycle needs to be evaluated in the context of these characteristics on repetition, standardization, reliability, and specifications.

Technology Transfer & Consciousness Development

We want to stand upon our own feet and look fair and square at the world -its good facts, bad facts, beauties, and ugliness. We have to see the world as it is and not be afraid of it.

We ought to stand up and look the world frankly in the face. We ought to make the best we can of the world, and it is not so good as we wish. After all, it will still be better than what these others have made of it in all these ages.

A good world needs knowledge, kindliness, and courage. It does not need a regretful hankering after the past or a fettering of the free intelligence by the words

MOSTINI Planet Next Level

uttered long ago by ignorant men.

It needs a fearless outlook and free intelligence. It needs hope for the future, not looking back all the time toward a past that is dead which we trust will be far surpassed by the future that our intelligence can create.

We have all, at some time, looked at the world around us and asked the same questions.

- Why does life have to be such a struggle?
- Why do we know so little about who we are and the purpose of our lives?
- Why is there so much conflict and suffering in a world of such beauty and with such riches?

I have experienced how we can tune our mind, and our consciousness, to other levels of reality and gain access to the unknown. Or at least to what is not widely known on Earth. I have realized that our minds and senses that are thinking and feeling for us are a series of energy fields, which use the body as a vehicle.

My consciousness is tuned to this dense world. A most important point to make is that, while on the same planet, a person's mind can tune in too many different wavelengths of knowledge and understanding. This is caused by variations in consciousness, perspective, and perception within humans.

In our daily lives, we even talk of people being on

MOSTINI Planet Next Level

different wavelengths because they think so differently and have attained different points of vibratory levels which their minds can easily access.

All this is essential background to what I believe is the truth behind the history of the human race. As previously mentioned, in the great struggle that enabled man to free himself from the animal, Nature was not the first shepherd of humanity but also its first enemy.

For man to free himself from this natural oppression, man has perpetually devoted himself to the conquest of the world. His essential aid was philosophy, science, and finally, technology. I discovered the real reactivity and reliable reactivities of the fundamental thermodynamic centrifugal law and microcrack extension stress stability centripetal law of the Universe.

These laws govern and control all actions in the Universe in all fields and domains, on a microscopic as well as a macroscopic scale with the stipulation of " .The fundamental thermodynamic force and microcrack extension stress stability (Entropy) are inversely proportional. The mass of the universe's entities system is proportional to the entropy of and inversely proportional to the fundamental thermodynamic force law.

Unfortunately, the reactivities of these laws are misinterpreted by scientists. Since the last century, they have given rude, incomplete, and inconsistent

MOSTINI Planet Next Level

interpretations of the understanding of astrophysics, geophysics, and mechanic.

However, humans have the impression that technology is evolving. In fact, the structure of the Universe, like an inverted plasma screen that , gives Mankind this impression and options. Of course, there is a certain revolution in technology applied to human needs (such as the means of communication, transport, and equipment). However, scientists are still far from knowing the real reactivities and functionalities of the Universe.

Gravity has been poorly explained since the beginning of time. Only 30% of research in astrophysics, geophysics, and medicine is reliable . So, I am incredibly sad to see billions of people in all the world lacking real gravitational knowledge. Mankind does not know the age of the Universe because they never understand the reactivities of planet formation.

The theory of the Big Bang as it is ascribed is not reliable and does not make sense scientifically. It seems to be mythologized. There are many enigmas and riddles all around us.

To illustrate taking account of actual tool (Newton and Einstein) for interpretation of Universe, many details of reactivity clarify the following for the human consciousness to know and learn:

1. Our star (Sun) is several times larger than all the

planets put together. According to current interpretations, gravity is the ratio of mass by attraction. It is also why planets do not strike and collide with the Sun (Stars) based on the ratio mass attraction theory.
2. Why a small apple has never been attracted by the Moon which is a huge satellite according to mass attraction.
3. How people could explain the Archimedeans Forces if gravity is related to mass proportion.

In this trend, the theory of the planetary disk is not understood by scientists. Many phenomena should be explained and revised with new evidence that is gathered reliably and proven scientifically.

Obviously, the physicochemical potential *Sigma ZG, Alpha & Omega Thermodynamic Matrix Sigma ZG, the Quantum Matrix Antigravity Sigma ZG, Geo-thermodynamic Sigma ZG,* are all four new scientific disciplines that are interconnected and are capable of explaining the reactivities of all the Universe.

They can also solve a proven scientific and reliable method all scientific enigma, riddle in microscopic scale as well in macroscopic scale.

Obviously, my research has led to the tangible realization of the first experiment on the work of the deformation of Universe by the principle of infinitely small

MOSTINI Planet Next Level

and infinitely large in mathematical correlation between the microscopic and macroscopic scale.

It has been realized in logic through the famous law, and formula $E=MC^{\wedge}2$ of *Einstein* followed by *Avocado, Boltzmann, Helmholtz free energy, Kepier William, Thomson Kelvin, NRNST,* and *RICCI (Italian Math)*.

My legitimate goal is to explain gravity, all universe reactivities by solving the scientific enigma and re-explain universe creation, the formation of the planet, stellar corona, the formation of clusters, supernovae, reactivities, and universe explanation in a reliable and real manner, with proven scientific evidence.

Moreover, I am about to explain radically, scientifically, and with mathematical demonstration, the mysteries of:

- The levers of Guatemala
- The Chiasm also known as the Chiasmus
- The Sinkhole
- The fairy circle of Namibia
- The legendary circle
- The methane bubbles of Canada in the Vermillion Lakes
- The series pink in Australia, (Chewing gum)
- The maelstrom,
- The Hum in Taos, New Mexico

MOSTINI Planet Next Level

- Clines and haloclines,
- Stellar corona
- Tsunamis
- Hurricanes
- Earthquakes
- Solar flares
- Tornadoes
- Chemoclines and more.

All of these are thermodynamic phenomena which are unexplained, due to the lack of understanding of the highest cosmic laws.

Also, I'm going to explain the importance of magnetism, the importance of radioactivity, and magnet reverse, which undergoes our planet actually.

Key Concepts of the Reality Approach of the Universe

Reality is grounded on the premise that human behavior is purposeful and originates within the individual rather than from external forces. All behavior is motivated by striving to fulfill basic the psychological needs one has of survival, belonging, power, freedom, and fun.

This approach focuses on solving problems, coping with the demands of reality in society and taking better control of one's life. By evaluating what we are doing and

MOSTINI Planet Next Level

what we want, we can achieve increased control of our lives

We perceive the world against the background of our needs and wants rather than as an objective reality. We create our own inner world. We are not locked into any one mode of behavior, although we must behave in the same way. Behavior is the attempt to control our perceptions and external world to fit our internal and personal world.

The choice theory teaches that the only person whose behavior we can control is our own.

How we feel is not controlled by others or events. We are much more in control of our lives than we realize. The choice theory explains how and why we make the choices that determine the course of our lives.

The choice theory assumes that all we can do from birth to death is behave. Everything we do can be understood within the context of total behavior, which is made up of four inseparable components: thinking, feeling, acting, and our physiology.

The reality of this book is to satisfy the curiosity of scientists, laymen, and philosophers. It is also, to provide educational and reliable information that stimulates the intellectual and emotional wellbeing of the readers to enrich their lives.

The aim is to show how mankind lives is affected

MOSTINI Planet Next Level

by the real universe gravitational knowledge reactivities with the legitimate purpose of reforming and updating education to improve personal evaluation at the global level.

Lost in This World

Lost in a world that does not look like me. In a universe where the true value of happiness is ignored

In this human society emptied of truth

I do not know who I am

What I know is that I do not know anything yet

I don't know if the orientation of this modern life confers happiness to humans, yet we have many mysteries, natural catastrophes, and injustice around us.

In this society empty of truth, I do not know who I am

What I know is that I do not know anything yet

I walk in the desert without seeing the horizon

I swim in a sea of darkness or accumulate all my fears with a burning desire to leave the cave of darkness and ignorance of the actions of the natural cosmic forces From this suffering, I am sick enough

I need to hold out

To encourage me to understand my environment,

MOSTINI Planet Next Level

highest cosmic laws, all that surrounds me, and all that is related to life in the Universe

I want to really understand the reactivities of the fundamental thermodynamic centrifugal force universe force and microcrack extension centripetal force stress stability

I want to understand how this Universe has been created

I want to understand this natural diversity and why I have been created

I want to understand the order of things

I want to understand the functioning mechanism of the Universe to deduce its properties, particularities, and prevent disaster.

I want to discover the naturals laws' reactivities, and our will has to be moved for the determinism of the natural laws.

The planet Earth is sick but it would be worse if we were all sick. My essence is the desire to be happy, live well, and act well by knowing the law that governs.

We believe in a better future powered by the people who have the courage to innovate in this world with their creativity. Despite the spectacular scientific advancements of the last century, there are many questions that researchers are still unable to answers.

MOSTINI Planet Next Level

Actually, mankind does not understand the reactivities and functioning's of naturals laws, the highest cosmic laws. There are still many questions that researchers are unable to answer for technical awakening and scientific progress.

Auguste Comte stated, *"The Ancestor rules the living being and their offspring."*

He translated it in a lapidary formula that *"All civilization is the fruit of the past."*

We cannot understand the present without constantly referring to the spiritual or intellectual inheritance of our ancestors to know the thoughts which shaped our world.

Knowing the thoughts of our predecessors, is not the need for learning but a need to find the thoughts that have shaped our World with geniuses like *Boltzmann, Riemann, Einstein, RICI, Euler, Nernst, Newton, Kepler, Avocado, Nicolas Tesla, Coulomb, Auguste Compte, Josammy, NRNST.*

The knowledge that you have created is the process of your thinking, and your love for knowledge engenders technological advances that are filled with light.

This is the reason to sing. The Sea that reveals itself to show murmurs is the need of love.

All these stars that fall right, it is the chance to you,

MOSTINI Planet Next Level

of you to make shine my City and to hear singing. And the horizons that light up in the corner of eyes is lot of desire to sing. The cloud that flies beyond the night is the dream that's dreams. And the sky is to big to hide our secrets.

Remember to look up at the stars and not down at your feet.

Try to make sense of what you see and wonder about makes the Universe exist. Be curious.

Chapter I
Mostini Planet

Nature always oppresses. For billions of years, humanity has been subject to its laws including the fundamental thermodynamic centrifugal force and microcrack extension stress centripetal stability. We have been at the mercy of tornadoes, earthquakes, and storms. All these natural disasters are only partially known and identified by us. They are not known in their fully comprehensive parameter of understanding.

They have often confused us and are partially strange to one's scientific understanding of the highest cosmic things and natural laws that apply in real life. We have other phenomena such as the natural levers of Guatemala, the Chasm, the Sinkholes, the fairy circle of Namibia, the methane bubbles of Canada, the Triangle of Bermuda, and the Lake Hillier in Australia (Chewing Gum), Maelstrom, the Hum in Tao, the Big Bang, and the universe expansion. These are just some thermodynamic phenomena, which are not known in their fully comprehensive reactivities.

I saw a new world arrive quickly. More scientific, more effective, yes. There were more treatments for old

MOSTINI Planet Next Level

diseases that were incredibly good. But it is a hard, cruel word.

And I saw a little girl, her eyes closed tightly, holding against her bosom the generous old world that she knew deep in her heart, could not stay, and she was holding it begging with me always.

Every day people awaken to a new reality of living in the grace of the present moment without remembering the past or projecting themselves into the future. An inner voice whispers to them let go of an old world that no longer corresponds to them, to enter a new wave that is certainly unknown but alive.

Mostini is a planet with three stars located in the Galaxy of Andromeda. Indeed, three solar suns revolve around Mostini Planet offering much more benefit to humans with the right thermodynamic conditions. A humans longevity is better here. On average they can live up to 300 years and have no diseases. Obviously, the detection of the exoplanets in multisystem stars represents the major problems in actual gravitational scientific theory (cosmology, astrophysics).

Unfortunately, a lot enigma and mysteries persist despite structural development in scientific equipment. Josammy Emporio Technology is the only technology that is capable of solving any puzzles or riddles and providing solutions to all the mysteries of the universe that we have.

MOSTINI Planet Next Level

I will do my best to address these universal questions here:

Question 1: What Is The Universe Made Of Or What Is The Composition Of Our Actual Universe?

Response: Ordinary matter has composition that is made up of protons, neutrons, and electrons. However, they only represent only 5% of the universe. The remaining 95% is made up of matter affected by the fundamental force and thermodynamic force centrifugal force. These exercise negative pressure in all universe with 29% of dark matter and 71% of dark energy actively exercising negative pressure in the entire universe.

MOSTINI Planet Next Level

Question 2: What are The Dimensions of the Universe and the Structures Parameters?

Response: Our universe is exceptionally large and uncountable. Our brain can't even imagine and realize the size of the universe, it is very huge. The universe is a hypersphere that is filled with the fundamental thermodynamic centrifugal force that increasingly spreads out from the center until the spherical sphere with three dimensions in a circumference that is curved in the 4D) surface force, spread out to the extremities of Holon universe.

MOSTINI Planet Next Level

Question 3: What is the Expansion Of the Universe And Why Is It so Easy Destroy?

Response: The fundamental Thermodynamic centrifugal force governs the universe and the fundamental thermodynamic force also impacts any matter, any substance and any entities to obtain stability and weak energy forms to reach the condition of lowest energy. The fundamental Thermodynamic force is hostile to any energy forms, and is always responsible for decreasing energy in the environment .These are seen as the lowest energy levels.

The fundamental thermodynamic centrifugal force affects all universe entities and particles in order to reduce energy to achieve a more stable state (Geo-thermodynamic). The fundamental thermodynamic centrifugal force is very hostile and is responsible for decreasing the energy in the environment. The entropy of the stars formation always destroys the stability of universe. The fundamental thermodynamic centrifugal force or the geo-thermodynamic Sigma ZG always favors all reactions and reactivities.

MOSTINI Planet Next Level

Question 4: Why Does Life Exist?

Response: Life is just the transformation of mineral chemistry to organic chemistry and this process is carried out by the fundamental thermodynamic forces in order to achieve the most stable states with lowest energy usage possible.

DNA is created from mineral elements that contain silica in order to achieve an organic transformation with lowest energy in the geographical environment. The reactivities of minerals elements is higher than organic systems. All organic systems undergo thermodynamic decay followed by putrefaction and volatility.

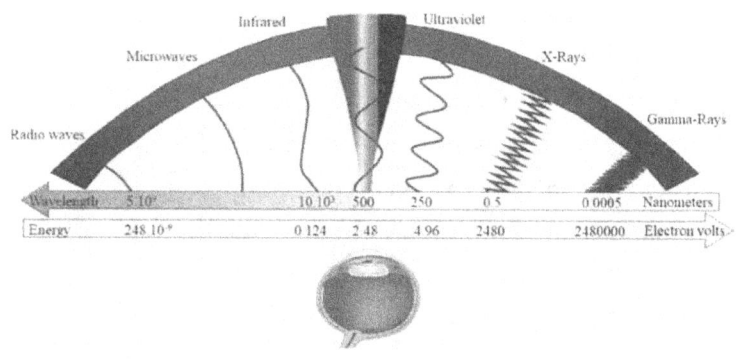

MOSTINI Planet Next Level

Question 5: Can Matter Escape Black Holes?

Response: The black holes represent the zone of the thermodynamic centrifugal barrier that is lowest in energy or has strict zero energy of the universe with zeta zero energy. The Black holes represent the zeta zero energy of the universe primordial before photo ionization of malice quantum disorder (Big Bang or photoionization).

Matter including photons can never escape black holes. The black holes demonstrate the evidence Big Crunch's reactivity in the permanent and continuity system of the reactivity of the universe. Black holes are thermodynamic barrier energies of the universe.

On the other hand, the Stars (Suns) and Supernovae are entropic barriers of energies. The fundamental thermodynamic force governs the universe. The fundamental thermodynamic centrifugal force and the microcrack extension stress stability centripetal force (Entropy) are inversely proportional to each other. This entropy is proportional to the mass system. The fundamental thermodynamic force centrifugal reactivities are inversely proportional to the mass entities systems.

MOSTINI Planet Next Level

Question 6: Why Are There Very Cold Zones and Very Hot Zones and Why Is the Universe Adiabatic?

Response: The reactivities zeta of spherical and all entire hyperspace is zero, all universe zeta is zero (z=0). And the reactivity of the entire universe is adiabatic. Therefore, the adiabatic process is a thermodynamic process in which no heat enters or leaves a system during expansion or compression of the fluid composing the system.

An example of an adiabatic process is the flow of air in the atmosphere; air expands and cools as it rises, and contracts and grows warmer as it descends. The temperature ΣT (cold-) + ΣT (heat+) = 0 conforming the condition of primordial universe zeta=0, Z=O, the hyperspace general temperature is zero.

Question 7: Do Entangled Particles Imprint the Universe Laws?

Response: The fundamental thermodynamic force always governs the universe. The entangled particles are affected by thermodynamics' affinity to the lowest energy.

Question 8: What is Time?

Response: Time is the reactivity's inversely between the fundamental thermodynamic centrifugal force

MOSTINI Planet Next Level

and the microcrack extension stress centripetal force stability which deforms the universe by the principle of zeta equilibrium Z=O reactivity's of the Quantum Matrix Antigravity sigma ZG and the physicochemical potential sigma ZG.

MOSTINI Planet Next Level

Question 9: Are We Alone in the Universe?

Response: Life is the transformation of thermodynamic mineral element towards organic element. Obviously, life is just the transformation of mineral chemistry to organic chemistry that is carried out by the fundamental thermodynamic in order to achieve the most stable states with less energy. DNA is created from mineral elements that contain silica to achieve an organic transformation with the lowest energy in the geographical environment.

In the universe, we find at least 200 billion Galaxies. Also, Galaxies are a huge collection of gas, dust and billions of stars and their systems, all held together by the reactivities of the fundamental thermodynamic force to the lowest energy. We are not alone because the reactivities of the fundamental thermodynamic is universal.

Question 10 - Do We Have A Distorted View Of The Universe?

Response: Scientists, philosophers, and laymen observe the world with a distorted view because they observe the universe from them inside. They are both observers and subjects of the observation. Their conclusion could be false by their subjectivity.

MOSTINI Planet Next Level

Question 11: What Are the Highest and the Greatest Universe Forces?

Response: We have 3 main forces which exert on the 3 dimensions of space that you should know about:

1. The fundamental thermodynamic centrifugal force
2. The microcrack extension stress stability centripetal force or the entropy centripetal force, which is also the gravitational centripetal attraction force, represented by heat energy, light energy, radiant energy, electromagnetic energy, chemical energy, and sound energy
3. The potential energy is a stored and centripetal force proportional to gravitational centripetal force. The fundamental thermodynamic force and entropy are inversely proportional

Question 12: What are the Highest Laws Reactivities of the Universe?

Response: In the universe, we have four laws of reactivities for the functioning of the universe.

1. The fundamental thermodynamic centrifugal force
2. The entropy energy centripetal force system
3. The gravitational centripetal force
4. The law of mass centripetal influence

Josammy Technology is the world technology

MOSTINI Planet Next Level

capable of explaining the reactivities of the planetary multisystem (Mostini Planet), Galactic system, hyperspace spherical universe 3 dimension curved in 4D, and solving all scientific enigma with the implication of the fundamental thermodynamic force centrifugal reactivities that are inversely proportional with the microcrack extension stress stability (universal entropy) centripetal force. Based on that, we have:

- Alpha & Omega Thermodynamic Sigma ZG - The Quantum Matrix Antigravity Sigma ZG
- The physicochemical potential Sigma ZG
- The Geo-Thermodynamic Sigma ZG are interdisciplinary technologies systems that focus on the reactivities of the universe including their reliable influence thermodynamic levitations causes.

Space Travel Adventure Inspection

Space travel is humanity's greatest adventure. It is the chance to explore the moon, the plants, and the stars. Giant rockets lift off with a roaring blast of orange flame. They climb into a blue sky, leaving a white trail. Then they speed out of sight in to space, where the sky is always black, and the stars are always bright and shinning.

The rocket may carry people on their way to conduct scientific experiments, or they may carry an artificial planet to explore a distant planet. Human beings

MOSTINI Planet Next Level

first set foot on the moon on July 16, 1969.

U.S. astronaut Neil A. Armstrong stepped out of the Apollo 11 part of the Apollo Lunar Module on 13:32 UTC. During the years since the space age began, many space travelers have been discovered.

The space age developed a huge industry called the aerospace industry to design and build space equipment. A new field of medicine called space medicine came into being to study the problems that come with living in space. Telephone calls, and television pictures are sent around the world by communications. Signals form navigation satellites that enable ship navigators. During the early years of the space age, success in space became a measure of a country's leadership in, science, engineering and national defense.

A planetary system is a set of gravitationally bound non stellar objects that are in or out of orbit around a star or a star's system. Therefore, in the universe we have found solar systems with one, three, and four Suns and more.

Obviously, a planet with three suns located around 340 light years from Earth in the constellation Centaurus, MOSTINI 2 is believed to be about 16 million years old. With a temperature of 850 kelvins (about 1070°F or 576°C) and weighing in an estimated four Jupiter masses.

It is also the coldest and least massive exoplanets. MOSTINI 2 is one the few exoplanets that have been

MOSTINI Planet Next Level

directly imaged, and it's the first one in such an interesting dynamical configuration. The reactivities of this planet with three stars or Suns causes much controversy reactivities on all the scientists' theories and knowledge until today. Actual scientist's knowledge is stipulating gravity as a ratio of mass that can't explain the reactivities of a single planet orbiting and revolving the three stars or Suns.

This single exoplanet's reactivities came to demonstrate once again the insufficiencies, the lakes and the limits of actual, recent scientist's knowledge and theories today because with their theory of mass ratio and proportion reactivities, this planet must absolutely crash on one of the three suns.

Obviously, Josammy Emporio 3D technology with Alpha & Omega technology sigma ZG, Quantum Matrix Antigravity Sigma ZG, Geo-thermodynamic Sigma ZG, and physicochemical potential sigma ZG technologies are capable to explains all universe reactivities with proven, reliable, and demonstrated scientific reactivities.

MOSTINI Planet Next Level

Alpha & Omega Thermodynamic Sigma ZG Matrix and Thermodynamic Function

Thermodynamics is a branch of physics that deals with heat, energy, compression and temperature, and their relation. In the universe, a chemical system is thermodynamically stable when it is at the lowest energy level.

Moreover, the chemical process usually occurs because they are thermodynamically favorable. Favorable means from high energy level to low energy level.

- The fundamental thermodynamic centrifugal force is refractory to the increase, and accumulation of energy such as: heat, light, radiant, electromagnetic, chemical, sound, nuclear, entropic, and all forms of energy.
- The fundamental thermodynamic force is refractory to any process likely to increase energy, and always maintain reactivities and reactions (chemical, biochemical, supramolecular assembly, and mechanical) to the lowest energy possible, in a more state or stage possible.
- The fundamental thermodynamic force impacts any matter or particles that have opposite charges to obtain stability and weak energy forms as possible.
- The fundamental thermodynamic force impacts the

particles in order to reduce energy to achieve a more stable state.
- When the fundamental thermodynamic solves an energy problem, another problem arises as a result of the first problem solved. To illustrate, we will cite the chain of planet creation, white dwarfs, pulsars, magnetars, supernovae reactivities, black holes, universe's entity aging, destruction and rust of material, birth and death, and expansion of the universe with the principle potential sigma ZG.
- The total energy between the particles with opposite charge may decrease as they approach each other.
- The fundamental thermodynamic force is hostile and refractory to the highest energy level. Therefore, when the fundamental thermodynamic force reduces or resolves one problem, the second problem occurs in consequence of the first problem solved in the goal and the legitimate objective of reducing all particles conforming to the stage and state of lowest energy.
- The thermodynamic fundamental centrifugal force can only orient an object or entities in all directions according to the state of the lowest energy of all reactions as well as mechanical reactivities including the electrical fields.
- The fundamental thermodynamic force governs the world.

MOSTINI Planet Next Level

Impact and Influence of Systems and 2nd Thermodynamic Laws in The Living System and Universal Entities

In the living system, all pathologies justify the states of low stability and high thermodynamic stability of chemical and biochemical reactions. The chemical process usually occurs because they are thermodynamically favorable. Favorable means from high energy to low energy, from less stable to more stable. Therefore, in the human body all biochemical reactions tend to be less stable with increase of time (aging) or when people are sick.

Obviously, all pathologies in the human body justify the state of the low energy. Finally, death is the loss of complexity; a loss of organization. Chemical reactions reach a stable level. A great loss of human entropy and disorder. However, living systems increase the entropy of their surroundings. Therefore, the cells create ordered structures from less organized starting materials.

Simpler molecules are ordered into the more complex structure of an amino acid, and amino acid are ordered into more complex polypeptide chains. At the organismal level, a complex and beautiful order results from the biological process that uses the simpler starting

MOSTINI Planet Next Level

materials. The silicon challenges the thermodynamic law energy in the human body and creates entropy and leads to the growth of cells.

The body's of living humans begin to fix silicon from fetal life to the brain, and ortho-silicic acid helps to run electric currents generated in the brain throughout the body. Silicon is incorporated in large molecules, and silicon is found in the cells, at the heart of the structure (centriole, nucleolus, mitochondria, and the cells membrane).

The fundamental thermodynamic force and entropy are inversely proportional. In the human body, ortho-silicic challenge the fundamental thermodynamic to stimulate the entropy of the body. The fundamental thermodynamic force governs and directs all metabolism in the living system. (Application of second thermodynamic law in the living systems). Silicon and ortho-silicic acid are both abundant in connective tissue, cartilage, skin, and the lymph nodes. The silicon and ortho-silicic acid challenge the fundamental thermodynamic force in the human body and help to regenerate tissues in wound, and promote multiplication, duplication, and replication of living cells.

This means ortho-silicic acid and silicon increase the physiological entropy in the human body to create organized complex structures, leading growth and cell regeneration. Obviously, a chemical system is thermodynamically stable when it is at low energy. This

MOSTINI Planet Next Level

means any living system (animals, plants, human body) is at the most stable level when death occurs.

In the universe, all entities obey the thermodynamic laws. The circular motion of the planets, stars, the Sun, Galaxies, pulsars, magnetars, solar systems, supernovae, all comply with the reactivities coordinated between the fundamental thermodynamic force and the microcrack extension stress stability (universe entropy).

In the living system metabolism reactivities, fatigue, hunger, sleeping, disease, finally death are all thermodynamic living system reactivities, to bring all the chemical systems to lowest energy with more stability. Obviously, death is the lowest thermodynamic stability of living system.

The round shape of celestial entities, aging of material, corrosion of materials, rusting of iron and materials, putrefaction of organic matter, the fading of the walls of buildings and houses, earthquakes, volcanoes, tsunamis, cyclones, magnetics inversion of planets, planetary corona, solar corona, the natural levers of Guatemala, the Chasm, the sinkholes, the fairy circle of Namibia, the methane bubbles of Canada, the triangle of Bermuda, the series pink lakes of Australia (Chewing gum), the maelstroms, the Hum in Tao are all thermodynamic reactivities to conform all reactivities and material to the lowest energy conforming the condition of

MOSTINI Planet Next Level

primordial universe to lowest energy.

Also, in the human's body sexual drive is justified by nocturnal ejaculation during sleep or the recurrent presence of supermatozoa in the urine when the male individual refrains from complying with this thermodynamic need to conform the body to the lowest energy. The evolution, adaptation, and improvement of style of life are all in line with the thermodynamic needs to the lowest energy, many species, both animal and plant, move to cooler region because of climate warming.

The migrations of animals and humans is the thermodynamic need in the goal to improve life in the condition of acting with the lowest energy level including the general evolution of mammals and all living systems. Also, the evolution and improvements of science and technology is the reactivities of the fundamental thermodynamic force to facilitate life and customs to lowest energies.

In the space, the reactivities of the fundamental thermodynamic force are high. When an astronaut stays in space, the organism loses the characteristics useful to its functioning on Earth because the space is filled up with highest fundamental thermodynamic centrifugal force reactivities or the highest amount of the fundamental thermodynamic force or dark energy. Therefore, the astronauts are under close surveillance and everything is

MOSTINI Planet Next Level

done to preserve physical conditions.

This is because of the high-level and amount of the fundamental thermodynamic force or dark energy in space, astronauts can lose around 20% of their bone mass within six months in space.

There is also the risk of developing kidney disease and all kinds of disease namely: Diabetes, Blood pressure Stroke, cancers and more if they stay for long time periods in space, for around 2 to 3 years. The human body deteriorates, as if it is aging rapidly. Muscles are quickly affected. The fundamental thermodynamic centrifugal force deteriorates all substances, entities, living systems entities.

Obviously, the Quantum Matrix Antigravity Sigma ZG is a phase of stability and continual interaction of the lowest energy. The Quantum Matrix Antigravity establishes the interaction level of strict thermodynamic stability to the lowest energy of the fundamental force and entropy.

This phase emanates irregular curvature deformation in search of kinetic stability. Moreover, this phase is representing and displaying the reactivities of the expansion of universe with the reactivities of the fundamental thermodynamic force.

MOSTINI Planet Next Level

Electricity Origin, Activities And Reactivities In Living Organisms (Humans, Animals, Plants)

Electricity is present in all the infinite universes as well as in antimicrobial cream and living organisms (humans, animals, plants). Therefore, in the human body, at the microscopic level, the cells that make up the human body, are bathed in a liquid containing all the nutrients, minerals and all the elements necessary for their proper functioning. The interior of the cell is separated from this extracellular medium by a lipid membrane called a plasma membrane, which forms a tight barrier between the intracellular and extracellular compartments.

Electrically charged atoms, called ions, can enter or leave the cell via transporters and membrane channels. Like an electric battery, it is the displacement of these ions between the inside and the outside of the cell and the resulting electronic imbalance which is at the origin of the electrical activity of the cells. This electrical activity is important for neurons, of course, because it participates in the coding and transfer of information, but also in other types of cells in the body.

For example: By contributing to cellular metabolism, the Antimicrobial skin cream, human body, and the universe present the same characteristic and properties. Therefore, the ortho-silicic acid, silicon are present the universe (sun, galaxies, planets and more) as well as in the antimicrobial skin cream and in the living organs (Humans, animals, and plants), they run electricity and control chemical reactions. The clay, Kaolin, contains silicon and living organs are the products of the universe. They are representing the creation of the cosmos. Living organ such as the silicon and orthosilicic acid and clay (covering the planet's surface) are the key to understanding the functioning of the universe.

Alpha & Omega Thermodynamic Sigma ZG

The fundamental thermodynamic force always reacts under the condition of reducing all the energy reactions (chemicals, biochemicals, mechanicals, electronics) and reactivities in the condition close to the reactivities of primordial universe which occurred before photoionization which is the first entropy of the universe. When the fundamental thermodynamic force regulates energies, we note, at the first time, the instability phase linked to the reactivities and co-reactivities that are inversely proportional between the fundamental thermodynamic centrifugal force and microcrack extension stress centripetal force stability (entropy).

Also, the fundamental thermodynamic reactivities phase process is the reactivity of all universals reactions

MOSTINI Planet Next Level

and reactivities (chemicals, biochemicals, mechanical, electrical, and supra molecular assemble) divided in two phases : Instability phase and stability phase,

1) Instability phase

This phase represents the beginning and the end of reaction. Obviously, all reactions and reactivities in the universe are instable at the beginning and the end For instance: The birth and death of Stars are always instable. In the universe, the instability of the beginning of reaction is called Alpha Thermodynamic sigma ZG. And the instability terminal or end of reaction is Omega Thermodynamic Sigma ZG. As a result, the reactivities instabilities at the beginning and the end of reaction is called Alpha & Omega Thermodynamic Sigma ZG.

2) Stability phase

The stability phase is the development, maturity, equilibrium to strict lowest energy, in the microscopic scale and macroscopic as well as in infinitely large and infinitely small. This stability and development phase reaction (The Quantum Antigravity Sigma ZG)

Also, the development, maturity, equilibrium, and stability phase reaction in the universe, at a microscopic scale and macroscopic scale as well as an infinitely large

MOSTINI Planet Next Level

and infinitely small level is called the Quantum Matrix Antigravity Sigma ZG. All the process reactivities are called Alpha & Omega thermodynamic matrix Sigma ZG.

Alpha is the instability at the beginning of the reaction. Omega is the instability at the end of the reaction, and we have the Alpha & Omega Thermodynamic matrix Sigma ZG (instability phase in the beginning and the end of reaction). The fundamental thermodynamic force always reacts under condition of reducing all energy reactions (chemicals, biochemicals, mechanicals, electronics) and reactivities in the condition close to reactivities of primordial universe occurred before photoionization, which is the first Entropy of the Universe. When the fundamental thermodynamic force regulates energies, we note at the first time, instability phase linked to the reactivities and co-reactivities inversely proportional between the fundamental thermodynamic centrifugal force and microcrack extension stress centripetal force stability (Entropy). We attend the phase of regularization thermodynamic which is stable and strictly stable.

Therefore, the fundamental thermodynamic force always reacts in two phases:

    a) The instability phase in search of thermodynamic stability. This phase is related first time of any reactions and reactivities in all the universe, for instance: The photoionization of the first phase of the Quantum Malice Disorder or the first universe photoionization (the disruption and inversion of the primordial universe), which generated the formation of stars, afterwards planet, solar systems, supernovae, pulsars, magnetars, clusters, and galaxies.

b) The stability development and stability

Also, the fundamental thermodynamic cycle is the reactivity of all universal reactions and reactivities (chemical, biochemical, mechanical, electrical, supramolecular assemble),

Chapter II

The Quantum Matrix Antigravity Sigma ZG Potential Sigma ZG

The Quantum Matrix Antigravity Sigma ZG is a phase of stability and continual interaction of the lowest energy. The Quantum Matrix Antigravity establishes the interaction level of strict thermodynamic stability to the lowest energy of the fundamental force and entropy. This phase emanates irregular curvature deformation in search of kinetic stability. Moreover, this phase is representing and displaying the reactivities of the expansion of the universe because this development stability reaction reactivities, the fundamental thermodynamic reactivities always favor reactions (chemicals, biochemicals, mechanicals, electronics) in infinitely large and infinitely small, from less stability and more to strict stable even with chemical reactivities like additional reaction, substitution, oxidoreduction, chelation, and supra-molecular assembles.

MOSTINI Planet Next Level

The energy equilibrium zeta of this phase is zeta=O

The Potential Sigma Z

A series can be represented in a compact form, summation or Sigma notation. The Greek capital letter: Σ is used to represent the sum. The series: 4+8+12+16+20+24 can be expressed read as:

$$\sum_{n=1}^{6} 4n$$

The sum is 84.

The representation sum sigma may be represented as follows:

$$(x+a)^n = \sum_{k=0}^{n} \binom{n}{k} x^k a^{n-k}$$

The potential sigma ZG represents the energy sum in the geographical environment. The potential Sigma ZG is a principal tool of energy reactivities or the sum amount of energy that use the fundamental Thermodynamic force and entropy. The fundamental Thermodynamic Force varies in the universe because the fundamental thermodynamic is inversely proportional to the mass of entities. This means the fundamental thermodynamic force varies. The more entities are huge, and the force thermodynamic is small; this means the highest attraction

MOSTINI Planet Next Level

Gravitation. The entropy is proportional to the mass entities. Therefore, potential Sigma ZG is the sum energy in the geographical environment G.

In the geographical area, the sum of energy should always conform to the condition of the universe primordial to the lowest energy. Obviously, the sigma ZG of stars represents the sum amount of energy of positrons, which is positive charge. The potential Sigma ZG of Stars represents the sum a—nt energy of total positrons in the geographical environment conforming to the lowest energy.

The potential Sigma ZG of Stars in the geographical environment is (+), this represents the sum number and number of positrons ejected during the solar flare. The solar flare is the thermodynamic reactivities to lowest reactivities of solar energy that leads to ejection of positrons in the Stars surrounding. The sum number of positrons in the geographical environment is called the potential sigma ZG of the Sun stars.

In the heart of all the universe's stars, we always basically assist the process transformation of hydrogen atom trough helium gas. This process of transformation reaction is the centrifugal thermodynamic process, which leads to a reaction from highest energy to lowest energy because helium is more stable than hydrogen atoms. The planetary potential of Sigma ZG is the sum energy always positive in the core. The planetary potential Sigma ZG is

MOSTINI Planet Next Level

always positive in the core, for instance: The potential Sigma ZG of our planet is positive because of presence of metal on the core iron and nickel. The stars 's potential Σ (positrons) are positive provided by Sun radiation and Solar flare moving toward Earth.

Planet Sigma Σ (iron/nickel) is positive. The solar flare is a tremendous sun explosion radiation released conforming of the fundamental thermodynamic force (from highest energy to lowest energy) in the universe all entities possess energy, which is heat, light, radiant, electromagnetic, chemical, nuclear, and electrical energy surrounding by appropriate electrical charge energy.

The total energy in the mass entities is the potential energy, the stored energy. There are two types of electric charges; positive and negative, (commonly) carried by protons and electrons respectively). Like charges with the same nature, repel each other due to naturally different ones attract each other. The current charge reactivities of the universe Stars are positive courant like the positrons of Stars and iron/nickel of the core of our planet.

The Sigma ZG of the telluric planet is positive like Earth planet with iron/nickel on the core. A telluric planet or rocky planet is a planet that composed primarily of silicate rocks or metals. Within the solar system, the terrestrial planets are the inner planet closest to the Sun, like Mercury, Venus, Earth, and Mars. They always have a

MOSTINI Planet Next Level

positive sigma ZG. Σ (ZG of the telluric planet) = Σ (+) in the geographical environment. Also, the four Jovian planets that exist are Jupiter, Saturn, Uranus, and Neptune.

A planet designated as Jovian is hence a gas giant, composed primarily of hydrogen and helium gas with varying degrees of heavier elements, The sigma ZG. Σ (ZG of Jovian planets) (+) in the geographical environment. They contain the same basic elements as a star. Jupiter and Saturn consist mostly of hydrogen and helium, with heavier elements making up 3 and 13 percent of the mass.

They are thought to consist of an outer layer of molecular hydrogen surrounding a layer of liquid metallic hydrogen, with a molten core. The layer of metallic hydrogen makes up the bulk of each planet and is referred to as "metallic" because the very large pressure turns hydrogen into electrical an electrical conductor.

The Jovian planets are located in the zone of high reactivities of the fundamental thermodynamic force of Dark energy, exercising a negative pressure on the surrounding and neutralizing and forming the planetary corona with Zeta $Z=O$, implying energy barrier of the planetary planet.

The Potential Energy ZG

The fundamental thermodynamic force, centrifugal force, and microcrack extension stress stability centripetal

MOSTINI Planet Next Level

forces are inversely proportional (Entropy). The chemical potential of a species is an energy that can be absorbed or released due to a change in the particle number of the given specie. The chemical potential of a species in a mixture is defined by the rate of change in the structure of free energy of a thermodynamic system. With respect to the changes in the number of atoms or molecules of the species that are added to the system.

Thus, it is the partial derivative of the free energy with respect reactivities of the fundamental thermodynamic and microcrack extension stability connected to the amount of all species, all other species concentrations in the mixture remaining constant. When temperature and pressure are held constant, the chemical potential is the molar Gibbs free energy. Because the fundamental thermodynamic force governs the universe, and the fundamental thermodynamic force and entropy are inversely proportional at equilibrium, the zeta reaction energy equals 0.

The zeta equals 0. In the universe, all interstellar round shape has energy zeta 0. When the several phases can coexist, each constituent will transfer from the in which its chemical potential is highest to the phase in which its chemical potential is the weakest (the fundamental thermodynamic force reactivities), until this chemical potential S equalizes. The zeta energy would then be z=0. In the universe, in equilibrium reactivities have z=0. For

MOSTINI Planet Next Level

instance, the round shape of stars, the planetary corona, the Solar Corona, natural satellite (our moon).

The Sigma ZG displays the stability and energy in the geographical environment. Obviously, the Thermodynamics is the form of energy that fills the entire universe and hyperspace in a uniform way. Equipped with negative pressure, it is a repetitive gravitational force that can explain the acceleration of the expansion of the universe. When it acts to reduce and favor all forms of reactions and interactions in a much more balanced situation the search for the zero value and zero entropy. which is the initial universe energy form.

$$\lim_{T \to 0} (\Delta S)_T = 0 \qquad \left(\frac{\partial E}{\partial S}\right)_{V, n_i} = 0$$

In the beginning, the reactivities or action and reaction of the universe were initiated by the process called, The Quantum Malice Disorder. The before creation one second, universe energy potential was zero.

$$\lim_{T \to 0} (\Delta S)_T = 0$$

The Zeta of the universe primordial was zero like an insulator. In a complex environment, the chemical of species influences and have an impact on the effect of the rest of the environment. The fundamental thermodynamic

MOSTINI Planet Next Level

and its legitimate force function exert an impact on the chemical potential and physical potential of all entities and living systems in the universe on the quest to favor all the universe reactions and reactivities.

The Physiochemical Potential Sigma ZG

The physicochemical potential testifies to the evolution and the situation of thermodynamic stability during the process phase reaction. The reactivities and co-reactivities of the fundamental thermodynamic centrifugal force inversely proportional with microcrack extension stress stability centripetal force entropy (Entropy), emanate curvature (normal, regular, multiform), Riemann, Boltzmann, Ricci, Einstein, in search of thermodynamic stability conforming universe to lowest energy.

The physicochemical potential Sigma ZG establishes the interaction level of thermodynamic (instability and stability) of chemical and physical reactivities. When the fundamental thermodynamic force regulates action or reaction, another problem arises which results from the first problem solved in the search for low stability.

We have the formations of the collisional and non-

MOSTINI Planet Next Level

non-collisional forces particles force. The potential physicochemical sigma is the topology curvature emanated by the co-reactivities between the fundamental thermodynamic force and entropy. We have topological curvature reactivities.

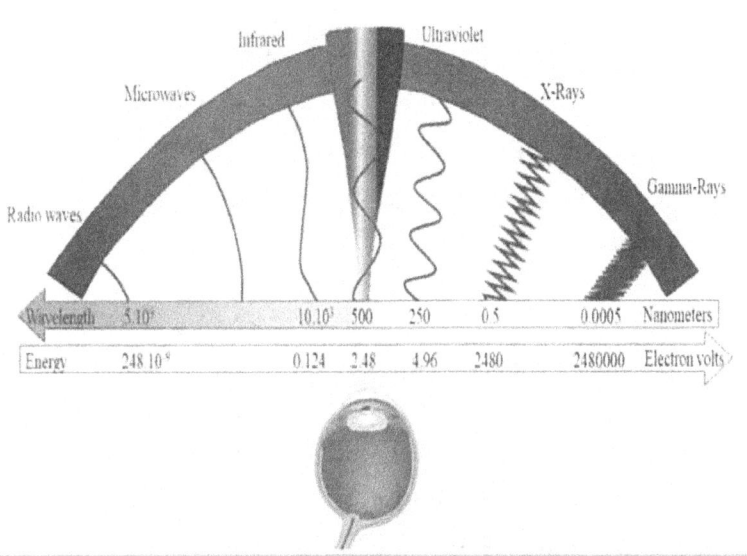

The figure displays the physicochemical potential of universe

The Geo-thermodynamic Sigma ZG

Environmental geography is the branch of

MOSTINI Planet Next Level

geography that describes and explains the spatial aspects of interactions between humans individuals or societies, and their natural environment. These interactions being called coupled human environment system. Therefore, the topic includes mountains, rivers, or cities.

Topological features like mountains affect weather mostly in the way the direction airs currents. The fundamental thermodynamic force governs the world, air force to rise over mountains. Moist air will cool as it rises. And then the clouds release the water, causing precipitation like rain.

Therefore, in the geographical environment, the fundamental thermodynamic force always favors all energies and reactivities to the lowest conforming energy level in a geographical area to the lowest energy as possible. The geo-thermodynamic Sigma ZG explains the behavior and paths of energy in a correlation of maintaining geographical are to lowest energy conforming the condition of Zeta=0 in the equilibrium states close to the Zeta of zero of the primordial universe, Zeta energy equal 0 was an energetic condition of the universe before the first photoionization (Malice Quantum Disorder).

MOSTINI Planet Next Level

Chapter III
Josammy Technology Application

The Thermodynamic Reactivities of Hydrogen

Atomic hydrogen is very reactive because it has the smallest atomic radius in the whole periodic table. In a reactivities series, the most reactive element is placed at the top and the least reactive element at the bottom. More reactive metals have a greater tendency to lose electrons and become positive ions.

The electron movement in the hydrogen atom may be modeled by the thermodynamic law. In the following process, emission and absorption of light during an electronic jump:

The electron can gain the energy it needs by absorbing light. If the electron jumps from the second energy down level to the first energy level, it must give off some energy by emitting light. The absorbs or emits in a

MOSTINI Planet Next Level

discrete package called a photon, and each photon has a definite energy. Only a photon with an energy of exactly 10.2 EV Can be absorbed or emitted must have a definite wavelength. 10.2 EV can be absorbed or emitted when the electron jumps between the energy levels and n-=2.

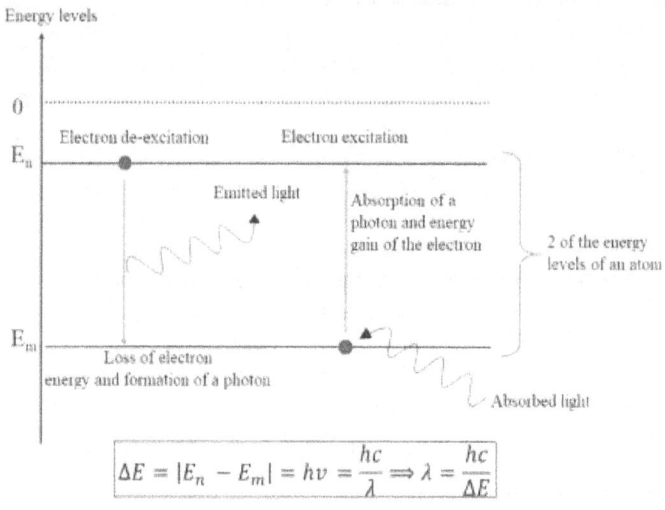

From a high-level of energy to a low energy level.

In the diagram, three different energy transitions have been represented.

Note that these transitions can be to a higher energy level (green) or a lower one (blue and orange). As thermodynamics governs the world, the electrons always go from a higher level to a lower level. In the fundamental

MOSTINI Planet Next Level

level the electrons are stable with the least energy very low energy level. The fundamental thermodynamic force always acts to keep the electrons at a fundamental low energy level.

For the electron to go to a higher energy level, it must gain energy (entropy). This energy corresponds to the energy difference between the starting level and the arrival level. The fundamental thermodynamic force and the entropy are inversely proportional. The electrons absorbs or emits in a discrete package celled photon.

From less stable to more stable

The fundamental thermodynamic force governs the world.

The Fundamental Thermodynamic Force Properties, Activities Law Reactivities And Description

Thermodynamics is the form of energy that fills the entire universe and hyperspace in a uniform way. Equipped with a negative pressure, it is a repetitive gravitational force that can explain the acceleration of the expansion of the universe, when it acts to reduce and favor all forms of reactions and interactions in a much more balanced situation, the search for the zero value and zero entropy.

MOSTINI Planet Next Level

$$\lim_{T \to 0} (\Delta S)_T = 0$$

This is the initial universe energy form.

$$\left(\frac{\partial E}{\partial S}\right)_{V,n_i} = 0$$

At the beginning, the reactivities or action and reaction of the universe were initiated by the process called photoionization and The Quantum Malice Disorder. The before creation one second, universe energy potential was zero.

$$\lim_{T \to 0} (\Delta S)_T = 0$$

The Reactivities Description Of the Universe Primordial Or Primitive

The fundamental thermodynamic force reacts like an elastic force with a legitimate rule to regulate and rehabilitate the universe in the condition of a primordial universe precisely always at the search of the zero, the regulation of zero entropy, corresponding with the entropy

MOSTINI Planet Next Level

of the universe before the creation. Obviously, before the creation of the universe (Malice Quantum Disorder), in the universe primitive, the entropy was zero, universe was an insulator with Temperature T=O, Pression P=O, Entropy S. In the universe the primordial as well as in the actual universe, the global Zeta energy is always zero, the Zeta z=0 although the structure, features, and reactivities are not the same. They are totally different.

The temperature approaches a limit which defines the notion of absolute zero. At the limit of absolute zero, a pure body in a perfect cosmos hyperspace in 3D dimension curved in 4D. Obviously, in the primordial era and environment (cosmos), the universe was like the crystalline, insulating surface or insulating area state has zero entropy. at T = 0 K, V p, S = O with absence of magnetism and radioactivity.

MOSTINI Planet Next Level

The Zeta Zero or Point Energy

The Zeta Zero-point energy of the universe is related to the reactivity of the fundamental Thermodynamic centrifugal force in the system or the geographical system. Is when the total charge in the system or geographic system equals zero or a high-level strict stability of thermodynamic stability. Therefore, there is no transformation, no chemical or biochemical reaction. We don't have any exchange the surroundings, and we reach the equilibrium state known as Zeta =0.

The fundamental Thermodynamic is reaching the equilibrium state Zeta =0. The fundamental Thermodynamic is reaching stability.

The entropy equals zero. This stability is obeying to the 3rd Thermodynamic law. The Zeta Zero or Zeta point Energy was zero before universe creation (Malice Quantum Disorder) and one second before The Malice Quantum Disorder.

$$\lim_{T \to 0} (\Delta S)_T = 0 \qquad \left(\frac{\partial E}{\partial S}\right)_{V, n_s} = 0$$

The chemical reaction between the pure crystalline phases that occurs at an absolute zero produces no change in entropy.

MOSTINI Planet Next Level

Nernst-Simon Statement: The entropy change resulting from any reversible isothermal transformation of a system tends to be zero as the temperature approaches zero

At the beginning, the reactivities or action and reaction of the universe were initiated by the process called, The Quantum Malice Disorder. Then before creation, one second, the universe's energy potential was zero.

The Universal Black Energy

$$\lim_{T \to 0} (\Delta S)_T = 0$$

The determination of the overall charge is made by measuring the zeta potential. The Third principle of Thermodynamic is associated with the descent to its fundamental quantum state of a system whose temperature approaches a limit, which defines the notion absolute zero.

Also known as the Nernst principle, the third of thermodynamic at limit absolute, a pure crystalline state has zero entropy. Therefore, at the beginning, before creation, the universe was like an insulator. Obviously, the primordial universe was a hyperspace spherical 3D curved in 4D with no entropy, a space insulator with no reaction, no transformation, no exchange just a space insulator era.

This experience justifies that the universe at the situation and the nature of the universe primordial at the

MOSTINI Planet Next Level

beginning before all creation. In this situation, the third principle of thermodynamics is associated with the descent to its fundamental quantum state of a system whose temperature approaches a limit, which defines the notion of absolute zero.

The entropy equals zero. The total charge of cream equals zero when the clay load on the antimicrobial cream is too high. There we do not have any exchange of the surroundings - we reach the equilibrium state.

The thermodynamic is reaching stability. The entropy equals zero. Therefore, the fundamental thermodynamic force is the universe's major composition with 30% dark matter and 70% dark energy. The universe primordial was insulator space.

MOSTINI Planet Next Level

Chapter IV
Barrier Energy and Thermodynamic Regulation

1. The Thermodynamic Matrix Sigma ZG Stability and Barrier Energy

The universal fundamental thermodynamic centrifugal force sometimes gives rise to kinetic stability (which is due to an energy barrier that is too high to overcome). Take the example of the diamond. Even if the fundamental thermodynamics law force favors the transformation into graphite, the energy required to achieve it is so high that we observe the kinetic stability.

In thermodynamics fields, reactivities, a thermodynamic system is in thermodynamic equilibrium when it is at the same time in thermal, mechanical and chemical equilibrium. The local state of a thermodynamic

MOSTINI Planet Next Level

equilibrium system is determined by the values of its intensive parameters, such as pressure or temperature.

More specifically, thermodynamic equilibrium is characterized by the minimum of a thermodynamic potential, such as Helmholtz free energy for constant temperature and volume systems or Gibbs free enthalpy for constant pressure and temperature systems.

The process leading to thermodynamic equilibrium is called thermalization. This process is thermodynamically stable, and the fundamental thermodynamics govern the universe. Planck declaration: For $T \to 0$, the entropy of any system in equilibrium approaches a constant independent of the other thermodynamic variables.

The theorem of the absolute inaccessibility of zero: there is no process capable of reducing a system's temperature to absolute zero in a finite number of steps. The entropy of any system is canceled. The potential in the fields of zero displaying straight line in the perfect situation of the fundamental Thermodynamic Force.

Obviously, in the universe, there are two types of barrier energies:

1) Thermodynamic centrifugal barrier energy, which is in the universe Black whole, and in the planetary system precisely in our planet the Bermuda triangle. The triangle features in the universe and displays the highest reactivities of centrifugal force to the lowest energy conforming the absolute zeta zero of the primordial universe. Because the fundamental thermodynamic centrifugal and microcrack extension stress stability centripetal force are inversely proportional, the Black hole has a round shape and cylinder.

2) The barrier energy Entropic gravitational attraction of centripetal force. This may be the supernovae with zeta =O, dwarves, magnetars, pulsars, stars.

2. Zeta non-Zero and The Quantum Matrix Antigravity

The potential of Zeta non-Zero of the universe is the co-reactivity process of all entities initiated just after the creation of the universe. The Matrix Thermodynamic First phase is initiated with the interaction between the fundamental Thermodynamic force and Entropy force particle providing current and instability led to the modification and deformation structure curvatures.

The Matrix Thermodynamic First instability phase emanates transformation topology with the effect of kinetic stability with collisional and non-collisional particles searching for Thermodynamic stability. Therefore, Kinetic stability is the consequence of the universe's undulatory nature.

MOSTINI Planet Next Level

When the fundamental Thermodynamic force regulates dark energy regulates action or reaction, another problem arises from the first problem solved in search of low stability. The second phase of the zeta non zero of the universe is the Quantum Metrix Antigravity in the first phase, to which the universe is the Quantum Metrix Antigravity in the first phase. The correlation between the fundamental Thermodynamic force reactivities reaches the strict to lowest energy stability level, and we have a new form of potential energy.

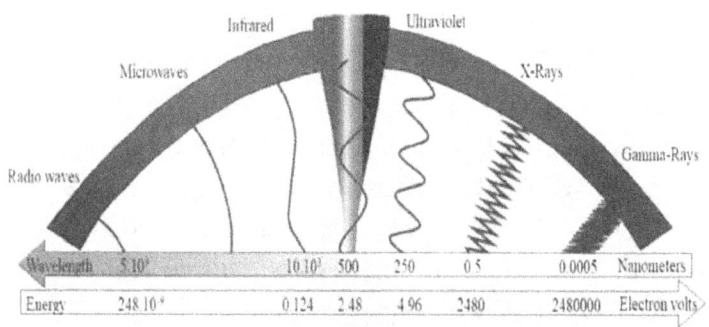

MOSTINI Planet Next Level

3. The Fundamental Thermodynamic Force Regulation in the Universe (Microcrack extension stress stability)

The fundamental Thermodynamic force governs the universe. The fundamental thermodynamic control all reactivities, core activities process of all entities in on a microscopic scale as well in the macroscopic scale to conform their energy in the condition primordial. Co-interaction and co-reactivities between the fundamental thermodynamic and microcrack extension stress stability in search of the lowest energy co-reactivities to conform to the condition primordial universe zeta zero. This dynamism of regulation of the fundamental thermodynamic forces exerts to bring back the universe in the primordial universe's physicochemical and electrical condition collides the entropy created by the Malice quantum Disorder. When the fundamental thermodynamic force regulates action or reaction, another problem arises from the first problem solved (there is the conservation of Entropy) in search of low stability or the condition primordial universe.

We have the formations of the Collisional and non Collisional force particles and a new form of energy, new nuclear electrical, new elementary microscopic.

MOSTINI Planet Next Level

The fundamental thermodynamic regulation force process is called "The Quantum Matrix Antigravity." The Quantum Metrix Antigravity regulates universe with several processes in all fields such as Neutralization regulation sigma ZG, electronic regulation Sigma ZG, the potential physicochemical sigma ZG, friction sigma ZG, Regulation Sigma ZG by influence thermodynamic, mechanic Sigma ZG, The potential sigma ZG.

4. Thermodynamic Regularization by Neutralization

In a complex environment, the chemical potential of species and potential physical influence impacts the rest of the environment on the species. The fundamental thermodynamic centrifugal exerts an impact on all entities' chemical potential and physical potential in the universe and living system in the universe, on the quest to favor all universal reaction and interaction. Thermodynamics force is always regulating all interaction and reaction on the quest to reach the absolute lowest energy. The fundamental Thermodynamic force always reacts to reach the state of lowest energy and has the tendency to orient itself to a low energy state or favor all chemical or biochemical reactions. In the universe and natural space, cations and anions always interfere with one another to produce much lower

MOSTINI Planet Next Level

energy. We can stipulate the neutralization reaction.

Moreover, two atoms separated by a very large distance from their diameters constitute two independent systems. By bringing them close enough to each other, a superposition of their respective orbitals takes place, which modifies the shape of the potential in which the electrons revolve. This results in a change of the orbitals themselves and, consequently, a variation of the electrons' energy. At this point, the two atoms form a system.

Atoms and particles in the universe exert a repulsive force on each other when the system's energy increases, as they get closer. Otherwise, the force is attractive. Because the thermodynamic force rules the world, and the fundamental thermodynamic force and microcrack extension stress stability are inversely proportional and interact. The existence of an attractive (Thermodynamic law) force is frequent, as evidenced by the extremely high number of different molecules and substances in condensed form, solid or liquid. Neutralization, chelation, reactions of assembled supramolecular, electrophilic, nucleophilic substitutions are thermodynamic reactions.

When an acid (H +) reacts with a base (OH-), water (H_2O) is obtained, a neutralization reaction is carried out, and zeta energy $Z=O$. The general equation is:

Acid + Hydroxide —> Water + Sal

On the contrary, we carry out purification methods

MOSTINI Planet Next Level

such as sublimation, which are Entropy process (Entropy means from high stability to less stability). We bring energy (heat energy) to create disorder and instability. We start from the solid-state (most stable) to finish with the gas state (unstable with Entropy disorder). Obviously, the solid state is a low state of energy. This means a high proportion of the fundamental thermodynamic force (more stable) and less Entropy (unstable). However, the gas is higher energy (less stability) than solid (most stability). This means the proportion of Entropy is higher while the fundamental thermodynamic force is lower. The fundamental thermodynamic and Entropy (microcrack extension stress stability) are always inversely proportional. However, Liquefaction (the passage of a gaseous body to a liquid state) and solidification (make into a hard or compact mass or change from liquid or gaseous to a solid.

The passage changes of matter that's a result in the production of a solid. The thermodynamics processes lead and favor chemical reaction from high energy states to states of low energy. The solidification can be carried out by thermodynamic processes such as cooling, pressure increase, crystallization, catalysis or by combining the phenomena that generate the reactions at lower energy. The liquefaction and solidification are the thermodynamic processes to lead reaction and reactivities from highest to lowest energy.

Neutralization is a thermodynamic regulation, so

MOSTINI Planet Next Level

acid (H+ strongly reactive) and base (OH- strongly reactive) react to form water and salt. A neutralization reaction occurs when an acid and a base react to form water and a salt and involve the combination of H+ and OH- to generate water. Obviously, the valence of an element is the number of hydrogen atoms with which it can combine. It can replace following the relation of the fundamental thermodynamic force and microcrack extension stress stability.

We call cohesion energy of a molecule, the energy induced by the fundamental thermodynamic force and microcrack extension minimum stress stability that must be available to dissociate the atoms that compose it and distance them sufficiently from each other so that they are no longer in interaction.

The cohesion energy is usually expressed as ev / molecule or kJ / mole. (1 kJ / mole = 1.037 ICT2 eV / atom). The study of valence bonds from fundamental equations of quantum physics has so far been successfully undertaken only in crystalline media and simple molecules. The fundamental force thermodynamics centrifugal forces explain the existing relationship with all existing chemical and biochemical bonds in the universe:

The fundamental thermodynamic centrifugal force law and microcrack extension stress stability centrifugal laws govern the world. The fundamental thermodynamic force governs the valence bonds in correlation with microcrack extension stress stability to obtain cohesion energy satisfying the rule of the thermodynamic rule of the lowest energy. The fundamental force thermodynamics and microcrack extension stress stability (Entropy) rules and governs the world. The fundamental Thermodynamics force and microcrack extension stress stability are inversely proportional.

There are three types of strong valence bonds, so named because of the high cohesion energy associated with them.

Those are:

- The Ionic Bond
- The Covalent (Or Homopolar) Bond
- The Metallic Connection

At the same time, there are weak valence bonds:

- Van Der Waals Binding
- The Dipole Bond
- The Hydrogen Bond.

Moreover, the fundamental Thermodynamic force refers to lead, and control of all chemical reaction: addition, chelation, supramolecular assemble, oxide-reduction, substitution, cation exchange, radical reaction

MOSTINI Planet Next Level

5. Regulation by the Principle of Action and Reaction

The action-reaction principle exhibits the co-interaction reactivity of the action and reaction under the influence of the fundamental thermodynamic centrifugal force and microcrack extension centripetal stress stability (Entropy). Obviously, in all life activities and for every action, there is an equal, opposite reaction and inversely proportional.

In the universe, the fundamental thermodynamic force regularizes all physical, electrical, chemical, and biochemical action by exerting the opposite and inversely proportional reaction to lead any action or movement (Entropy) to low energy.

The fundamental thermodynamic force governs the world. The equal and opposite reaction means that a force of the same intensity will be applied but at 180 degrees of the initial force. Let's use a very simple example. Suppose you put both your hands on the wall of your room, and you push with all your strength, 15 N, for example. The wall will not move because it is too solid, but you will go back. As farfetched as it may sound, the wall pushes you. Indeed, when you push on the wall at 15 N, there is an equal and opposite reaction. In conclusion, the fundamental thermodynamics force and the entropy (macrocrack

MOSTINI Planet Next Level

extension stress stability) are inversely proportional. This reaction is that the wall grows on your hands with a force of 15 N. This is also how our legs make energy for walking. The floor pushes our feet forward and moves us forward. If the floor is too slippery, our feet cannot push on the ground, and we slide.

Moreover, if the reaction is equal and opposite, why it is not canceled by the action. Of course, you know that two equal and opposite forces cancel each other out, but in this case, they are inversely proportional to each other. It is not the case because these two forces do not have the same points of application. We cannot add them up, and they do not cancel each other out. Fundamental thermodynamic force and microcrack extension stress stability (Entropy) are inversely proportional. The fundamental thermodynamic force always acts to favor all the chemical, biochemical reaction and mechanical force; for instance, have you ever skated?

If so, you must have realized that your body moves back by pushing the band with your hands. Why exerting a force on one wall makes us move back in the opposite direction? The answer to this question is related to the action-reaction" the action-reaction under the influence. The implication of the fundamental thermodynamic force and macrocrack extension stress stability (Entropy) principle is inversely proportional. If an object A (Entropy force) applies a force on an object B, then object B

MOSTINI Planet Next Level

(Thermodynamic reactivity) will apply a force equal and opposite to the object A principle, which is Newton's third law. This principle can even explain how to walk. Thermodynamic force governs the world.

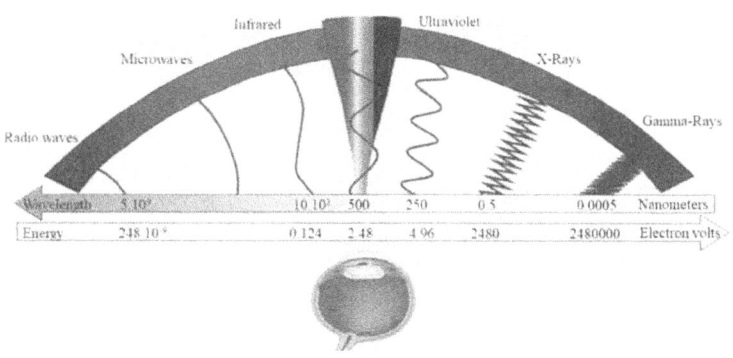

The fundamental thermodynamic force and microcrack stress stability (Entropy) are opposite and inversely proportional. This diagram shows the coordinated activity of the fundamental thermodynamics force, and microcrack extension stress stability (Entropy) in continual, perpetual correlation , and this these co-activities of the fundamental thermodynamic force and Entropy governs the world, all the activities of the universe (the mechanical movements of all, the living beings, any material object perceptible and not perceptible, existential obey to this law. The fundamental Thermodynamics governs the world. We may practically display the correlation between the

MOSTINI Planet Next Level

fundamental thermodynamic and macrocrack extension stress stability, which are inversely proportional. This perpetual correlation may be called the Quantum Metrix Antigravity.

6. Electronic Regulation Sigma ZG

6.a. The Fundamental Thermodynamic Force and the Particle In Electric Field

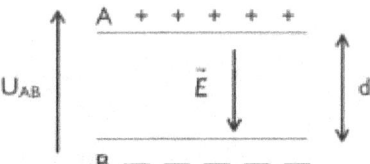

A charge q (C) undergoes a force De = I q l x E = l q I x U / d if it is placed between two charged plates, distant from d (m) and subjected to a voltage U (V) and exerting a field E (C / N). The particle placed in an electric field moves towards a decreasing potential, which justifies the action, and the fundamental participation force thermodynamic force directs the charged particle towards a more stable potential requiring less energy. The fundamental thermodynamic force governs the world and favors chemical, biochemical reactions, and electrical, mechanical interaction.

In the common sense, energy refers to anything that makes it possible to do work, to produce heat, to create light, to modify chemical species, to generate movement. Therefore, energy is a physical, numerical quantity, which has no general character because associated with a concrete situation. It is neither fluid nor a substance. We speak of movement energy, electrical energy, chemical energy, thermal energy, nuclear energy, hydraulic energy, wind power.

In the following process, emission and absorption of light during an electronic jump:

MOSTINI Planet Next Level

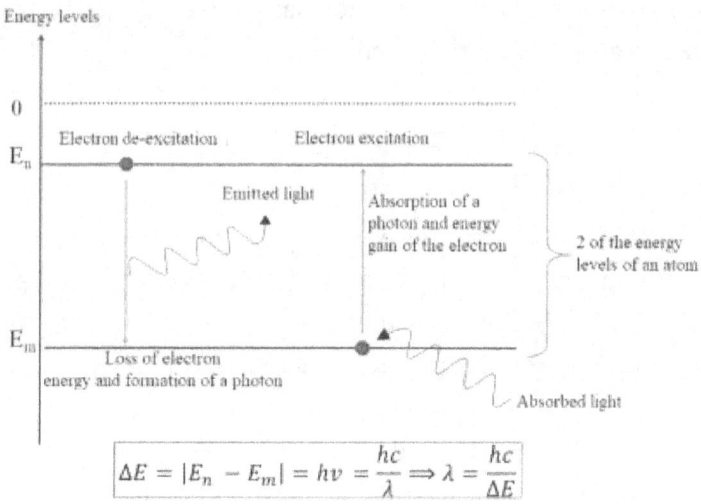

The thermodynamic force always acts to keep the electrons at a fundamental low energy level.

For the electron to go to a higher energy level, it must gain energy (entropy). This energy corresponds to the energy difference between the starting level and the arrival level. The fundamental thermodynamic force and Entropy are inversely proportional.

This energy, denoted by ΔE, is therefore equal, in our example, to the difference of the two energies

MOSTINI Planet Next Level

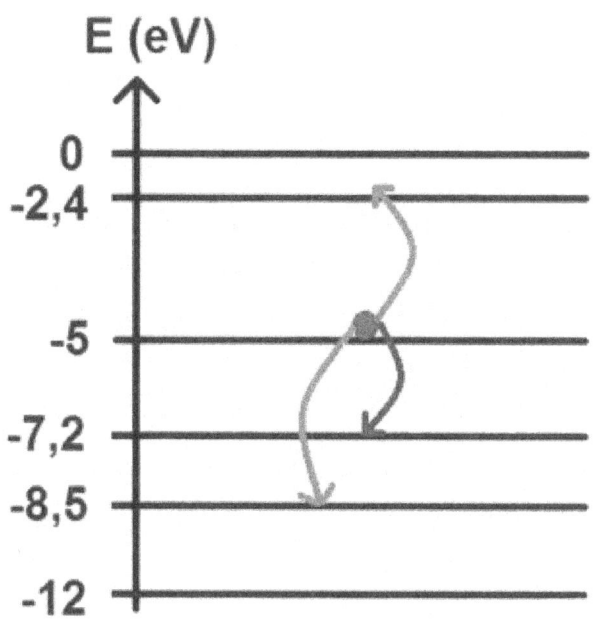

The energy gain can be made if the atom absorbs light radiation of frequency or wavelength I (in the spectrum of the light illuminating the atom). We will have a dark line on a bright background, and the spectrum of the received light will be a spectrum of absorption lines).

The emission of light can be done by losing the electron's energy, if the electron passes from an excited level to a level of lower energy: bright line on a dark background, and we will have a spectrum of lines of emission.

MOSTINI Planet Next Level

In the universe, the state of lowest energy is comparable with the fundamental level of atom energy. Therefore, thermodynamics is a fundamental energy force in the universe. The thermodynamic force governs the universe.

The diagram fundamental state is the perfect and stable level lowest energy. At the limit to absolute zero. Also known as Nernst's principle, the third principle of thermodynamics is:

The third principle of thermodynamics

At the limit of absolute zero, a pure body in a perfect crystalline state has zero entropy.

$$\text{at } T = 0 \text{ K}, \forall p, S = 0$$

$$\left(\frac{\partial E}{\partial S}\right)_{V, n_i} = 0 \qquad \lim_{T \to 0} (\Delta S)_T = 0$$

Thermodynamics governs the universe. In the relation System and Universe, Thermodynamics aims to study the exchange of energy and matter between parts of the Universe. The Universe is the whole of space and matter accessible to our knowledge.

MOSTINI Planet Next Level

6.b. Geographical Area Regulation or Geo-thermodynamic

When separating the positive and negative charges, a cloud gives rise to lightning; the energy created results from thermodynamic instability. Therefore, fundamental geo thermodynamic force and the macrocrack extension stress stability always regulates the electric charges or the electromagnetic energy when there is a disturbance of the charges. To regulate the geographical environments in the conditions of the primordial universe with zero energy zeta—O.

The turbulence of a storm with its upward and downward currents represents the ideal environment for separating electrical charges. The fundamental geo-thermodynamic force and macrocrack extension stress stability favors the location of negative charge concentrations at the base of the clouds, while the positive charges begin in the upper parts

This phenomenon allows the electric field to form and fill up between the cloud and the soil by the principle of geo thermodynamic and microcrack extension stress stability of the solar system (the sun and planet relationship in the levitation, interaction, and attraction reactions)

Similar charges repel each other, and the opposing charges attract each other. The negative charges begin to

MOSTINI Planet Next Level

spread near the cloud base, much like a shadow. Positive charges tend to focus on objects and heights. A cloud-ground eclair begins when a negative charge makes its way to the ground. This phenomenon tracer bond can go up to 100m each. This leap tracer can go in several directions under the thermodynamic stability control

In response to the negative charge from the base of a cloud, positive charge currents begin to move upward from the ground. This is an ascending tracer. The fundamental thermodynamic force and macrocrack extension stress stability always regulates the positive and negative charges to reach the conditions of the primordial universe (Energy zero, Zeta= 0, t=0, S=0)

$$\lim_{T \to 0} (\Delta S)_T = 0$$

$$\left(\frac{\partial E}{\partial S}\right)_{V,n_j} = 0$$

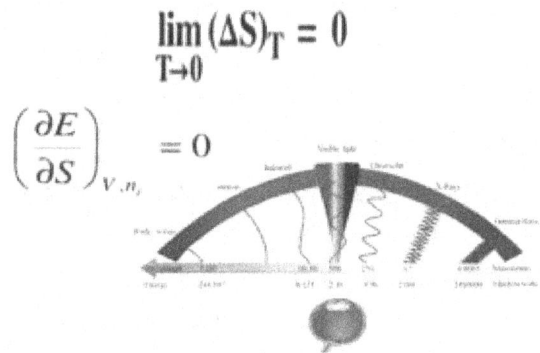

MOSTINI Planet Next Level

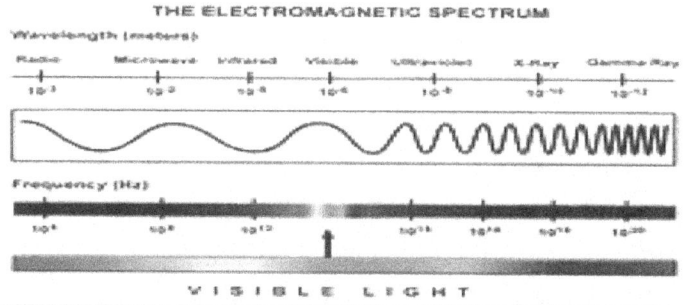

The graph shows and exhibits the continual and perpetual correlation between the fundamental thermodynamic and microcrack extension stress stability by an inversely proportional relation, which justifies that all the entities in their universal interactions, properties, and activities always obey the entropic and thermodynamic law are inversely proportional in the universe. The fundamental Thermodynamic Force govern the universe.

The Quantum Metrix Antigravity is the regulation, co-interaction and co-reactivities between the fundamental thermodynamic and microcrack extension stress stability in the search of co-reactivities of lowest energy possible to conform to the condition primordial universe zeta zero. This dynamism of regulation of the fundamental thermodynamic forces exerts to bring back the universe in the primordial universe's physicochemical and electrical condition collides with the entropy created by the Malice

MOSTINI Planet Next Level

Quantum Disorder. When the fundamental thermodynamic force regulates action or reaction, another problem arises which results from the first problem solved (there is the conservation of Entropy) in search of low stability or the condition primordial universe. We have the formations of the Collisional and non-Collisional force particles and a new form of energy, new nuclear electrical, new elementary microscopic, new macroscopic particles, and protoplanetary formation disks. In the conditions of co-interaction and inversely proportional of the fundamental thermodynamic and microcrack extension stress (entropy) in a constant and permanent regulation giving electromagnetic waves. The protons have a high ionization density.

$$E=MC^2 \quad \text{Gravitational waves}$$

One of the most astonishing consequences of the interaction and co-interaction between the fundamental thermodynamic and microcrack extension stress stability, general relativity is the presence in the universe of gravitational waves. Indeed, a deformation of space-time, induced by the presence of entities and body, does not disappear like that (conservation and transformation of energy). It propagates, at the speed of light, like a sound or light wave. In a mutual electromagnetic and electrostatic thermodynamic levitation's activities, the Sun and planets generate gravitational waves that are far too weak to be detected. However, very violent events involving very

MOSTINI Planet Next Level

massive bodies in the universe can cause waves of sufficient intensity to be detected.

When a light ray propagates in a gravitational field, it spends some of its energy to "go up" in this field. This loss of energy then manifests itself by a shift of the electromagnetic spectrum towards the red, i.e., at a weakening of the light ray. solar flare outburst of energy is always associated with areas of the solar magnetic field. It can also be interpreted using the principle of equivalence: the fundamental thermodynamic force and microcrack extension stress stability generate the gravitational field in which the photon is immersed is equivalent to a uniformly accelerated reference away from the observer. Electromagnetic waves' presence is related to the reactivity of the fundamental thermodynamics and macrocrack extension stress stability, which are inversely proportional forces.

MOSTINI Planet Next Level

The Quantum Metrix Antigravity

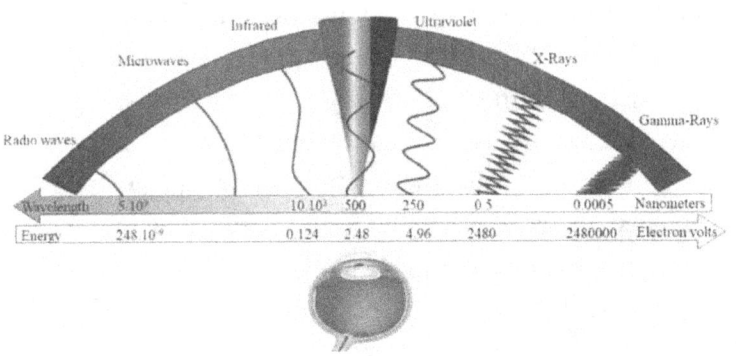

7. The Potential Physicochemical Sigma ZG Regularization and Physical Impact

When the fundamental thermodynamic force regulates a reaction, another problem arises, which results from the consequence of the first problem solved in search of low stability or the condition of thermodynamic stability. We have the formation of the collisional and non-collisional forces particles; the Quantum Matrix Antigravity establishes the interaction level of strict thermodynamic stability to the lowest energy between the fundamental thermodynamic force microcrack extension stress stability. Therefore, The Quantum Matrix Antigravity

MOSTINI Planet Next Level

generates topological curvature deformations with Normal curvature, regular, multiform, Riemann, Boltzmann, Ricci) in search of stability it proceeds kinetics speed Movement of stability. The fundamental (thermodynamic) force of the universe exerts a negative pressure in all universal reactions and universal entities to bring them back to the primordial universe's physical and chemical conditions.

Because the fundamental thermodynamic and macrocrack stress stability are inversely proportional, the kinetic moment, gravitational moment, electric moment, magnetic moments are induced modeled, initiated for the search of physical and chemical condition primordial universe. The physicochemical regulation potential in the universe and our planet could be observed tangibly because of co-reactivity between the fundamental thermodynamic force and Entropy. The fundamental Thermodynamic force and Entropy are inversely proportional. For instance: We have hurricanes, tornadoes, earthquakes and storms, the natural levers of Guatemala, the Chasm, the Sinkholes, the fairy circle of

Namibia, the legendary fairy circle, Canada's methane bubble, The triangle of Bermuda, the series of pink lakes in Australia (Chewing gum), The maelstroms, Hum in Tao.

The graph showing that the two periods of the vegetation of the wheat are attached, by an inversely

MOSTINI Planet Next Level

proportional relation which justifies that relation, which justifies that all the entities in their universal interactions, their properties and their activities always obey the entropic and thermodynamic law are inversely proportional in the universe. Thermodynamics spins and runs the universe.

MOSTINI Planet Next Level

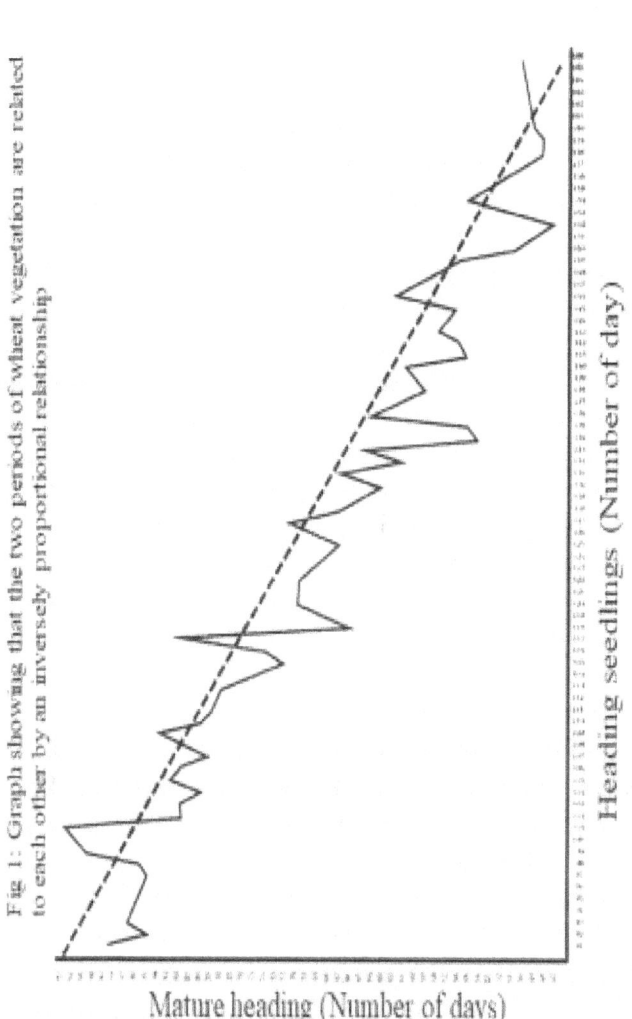

Fig 1: Graph showing that the two periods of wheat vegetation are related to each other by an inversely proportional relationship

MOSTINI Planet Next Level

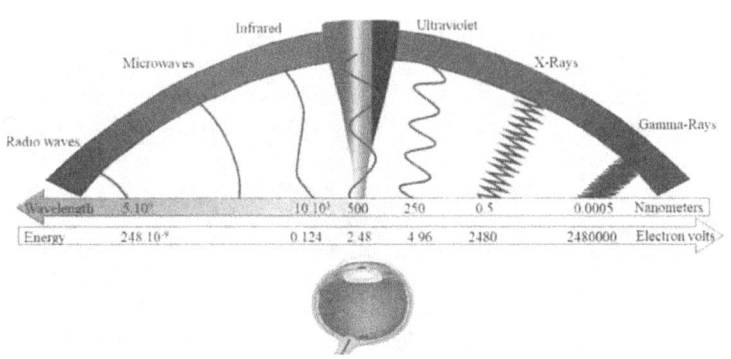

8. Thermodynamic Regulation by Friction Sigma ZG

When two bodies (insulators tools) are rubbed against each other, one pulls electrons out of the other. The body that has an excess of electrons is negatively charged. The same reactivities apply to the infinitely large and infinitely small macroscopic scale and macroscopic scale.

The body or that has lost electrons is positively charged. For example: the glass rubbed against the wool is positively charged because the wool pulls electrons. After a while, like 30 minutes, the negative charged and positively will attract in the geo-thermodynamic condition. The fundamental thermodynamic force and microcrack stress stability interact to neutralize charge in the geographic area.

MOSTINI Planet Next Level

Obviously, in this area or system. The total charges are neutralized; we notice the geographic situation of zero energy.

You all observed, one day, while painting, that your hair was attracted by the comb. The same attraction phenomenon appears when you unpack an item wrapped in cellophane.

The comb, the cellophane sheet electrified.

These phenomena are known since ancient times. Twenty-six centuries BC (Joliot-Curie), Thalès de Milet amused himself by rubbing yellow amber with a cat's skin. When two bodies are rubbed against each other, one pulls electrons out of the other.

The body that has an excess of electrons is negatively charged.

The body that has lost electrons is positively charged. For example: the glass rubbed against the wool is positively charged because the wool pulls electrons.

How to know when one rubs two bodies, the one that will pull electrons to the other?.

Example: the glass rubbed against the wool is positively charged because the wool pulls electrons

How to know when one rubs two bodies, the one that will pull electrons to the other?

MOSTINI Planet Next Level

See triboelectric list: (example: rabbit skin, glass, wool, cat skin, cotton, ebonite, plexiglass, nylon) Electrification by contact.

When a negative body touches a neutral body, electrons can pass over the neutral body, which becomes negative. When a positive body touches a neutral body, it attracts electrons from the neutral body which then becomes positive. We may have: Geo-thermodynamic works in the same frameworks in the fields of microscopic and macroscopic entities.

MOSTINI Planet Next Level

9. The Fundamental Thermodynamic Force Regulation sigma Z G by Influence or levitation

The electrostatic pendulum consists of a luminous ball (elder, expanded polystyrene) covered with a conductive layer (a sheet of aluminum, graphite) suspended from a support by a wire.

At the approach of an electrified wand of the pendulum, the ball is attracted by the wand. After contact with the stem, the ball is pushed back by electrical loads

It is said that the bodies are electrified because of the presence of very small electrical charges.

Suppose we electrify an electrostatic pendulum by contact with a charged rod, and we approach successively other electrified bars. In that case, we see that the ball of the pendulum is either attracted or repulsed by the different bars. In this process, the fundamental Thermodynamic force and quantum Metrix Antigravity regulate the macroscopic scale as well in the microscopic scale.

Moreover, a leaf electroscope consists of a metal rod supporting two narrow and very thin gold or aluminum sheets. The set is placed in a transparent and insulating enclosure (glass).

MOSTINI Planet Next Level

When an electrified stick is approached (without touching it), the electroscope leaves move apart. If we move away from the wand, the leaves fall back.

The leaves repel because they are electrified under the influence of the wand. In the previous experiment, the electroscope, without contact with the rod, could not be loaded. It is globally neutral. If the slats have moved apart, it is because electric charges have moved inside the metal. These are free electrons. If the rod is negative, it repels the free electrons of the electroscope. These electrons are found in excess in the lamellae that become negative and repel. (see diagram opposite).

If the wand is positive, it attracts the free electrons from the electroscope. These electrons are found in default in the lamellae that become positive and repel. On the microscopic scale as in the macroscopic scale, the fundamental thermodynamic stability would tend to favor all states to the lowest energy possible. Sometimes, it takes a lot of energy to move from one to another. This gives rise to kinetics stability (which is due to energy to overcome).

10.Geo-Thermodynamic Sigma ZG Riemann deformation Regulation

The Quantum Metrix Antigravity is the regulation, co-interaction, and co-reactivities between the fundamental

MOSTINI Planet Next Level

thermodynamic and microcrack extension stress stability in the search of the lowest energy possible co-reactivities to conform to the condition primordial universe zeta zero. This dynamism of regulation of the fundamental thermodynamic forces exerts to bring back the universe in the primordial universe's physicochemical and electrical condition collides with the entropy created by the Malice quantum Disorder.

When the fundamental thermodynamic force regulates action or reaction, another problem arises which results from the first problem solved (there is the conservation of Entropy) in search of low stability or the condition primordial universe. We have the formations of the Collisional and non-Collisional force particles and a new form of energy. In the macroscopic as well as in the microscopic scale. The potential Sigma ZG always impacts in the geographic environment in order to reach the state of lowest energy.

Moreover, the fundamental thermodynamic force evaluates the energies values of celestial universal entities (Macroscopic scale and microscopic scale) to conform their energies in the condition of lowest energies in the geographical environment.

When the value of potential Sigma ZG governs the particles, that the atoms are excited with an increase of energy in a geographical, the potential Sigma ZG impact

MOSTINI Planet Next Level

and assess the situation by evaluating their energies in the geographical, qualitative, and quantitative orientation to obtain the values of lowest energies in the universe the rotation of planets, the propagation of universe, the formation of Galaxies, Solar Systems, and all orientation of universe entities, all activities manifest themselves by affinity and repulsion of stability reactivities to lowest energies.

For instance: universe propagation, solar systems rotation reactivities, the round shape of celestial entities, the configuration of Planets, Whites dwarves, Black holes. In the universe, all entities (Microscopic and Macroscopic scale) deform with a rapport of Mathematic and energy evaluation. The physicochemical potential Sigma ZG always evaluate, detect, identify, and classify the potential of all entities and reaction to conform in strict stability condition to lowest energy. The energy evaluation emanates the deformation curvature and orientation of gravitational wave, X-Ray, Gamma Ray, Radio waves, and UV ultraviolet. The structure is geometric, most often metric, Reimanienne.

Reimanienne's Hypothesis Zeta Function

$$\zeta(s) = \sum_{n=1}^{\infty} \frac{1}{n^s} = \frac{1}{1^s} + \frac{1}{2^s} + \frac{1}{3^s} + \cdots$$

MOSTINI Planet Next Level

Leonhard Euler introduces it (without giving him a name) only for real values of the argument (but also for s = 1). In his theory of special relativity, in addition to the dilatation.

$$\zeta(s) = \prod_{p\ \text{prime}} \frac{1}{1-p^{-s}} = \frac{1}{1-2^{-s}} \cdot \frac{1}{1-3^{-s}} \cdot \frac{1}{1-5^{-s}} \cdot \frac{1}{1-7^{-s}} \cdot \frac{1}{1-11^{-s}} \cdots$$

The statistics connection, among others, with his solution of the Basel, which react to the spacings between the different zero of Reiman's function, is exactly the same as the statistics that govern the theoretical random atom's energy levels.

In relation to the fundamental thermodynamic force and microcrack extension stress stability, if we build an atom with random bricks, measure energy levels, and find the statistics of the state of Reiman. We have the connection and correlation between quantum mechanics and prime numbers. The differences between the zero of Reiman are the same as the statistics of the energy levels, which exist at a random level.

The time and the contraction of lengths, Einstein has brought to light a new form of energy, often called "resting energy,' or " mass energy. " As we know, a

MOSTINI Planet Next Level

moving entity or body has some kinetic energy. But Einstein shows that he also has one when he is at rest, it is given by the famous relation:

$E = MC^2$ where M is written as the mass at rest. How to understand this relationship? It shows that every mass is equivalent to one energy and vice versa. If we could disintegrate an object and turn it into pure energy, we would get a quantity of energy equal to MC^2 - the space-time metric and thermodynamic

The principle of equivalence is the key to understanding general relativity and the origin of the completely new approach to gravitation. The principle of equivalence gives us a way of describing gravitation other than by force (Newtonian approaches) by directly using the notion of a uniformly accelerated landmark.

Where the infinite product bears on all prime numbers p but does not necessarily converge, also, Euler gives proof of this formula for the case s = 1. The Thermodynamic relation and Entropy spins, govern, run and controls the universe, and are inversely proportional. We may also display, Boltzmann curvature(S=klogw), Riemann curvature, Ricci curvature.

MOSTINI Planet Next Level

11. The Mechanical Sigma ZG Thermodynamic Regulation

The fundamental thermodynamic force mechanic Sigma ZG Regulation and projectile

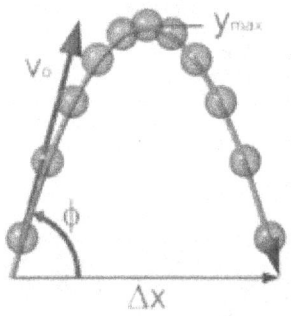

A projectile launched from the ground always falls to the ground because of the fundamental thermodynamic force's reaction, which brings back the projectile to an initial state of weak and low energy at the end of its movement. Since the fundamental thermodynamics and Entropy (the initial velocity) are inversely proportional. The speed of arrival of the projectile is identical to those he had initially (assuming that the soil at the end of the movement is neither higher nor lower than the ground at the start) because the projectile motion is symmetrical at its midpoint. The fundamental thermodynamic force governs the world.

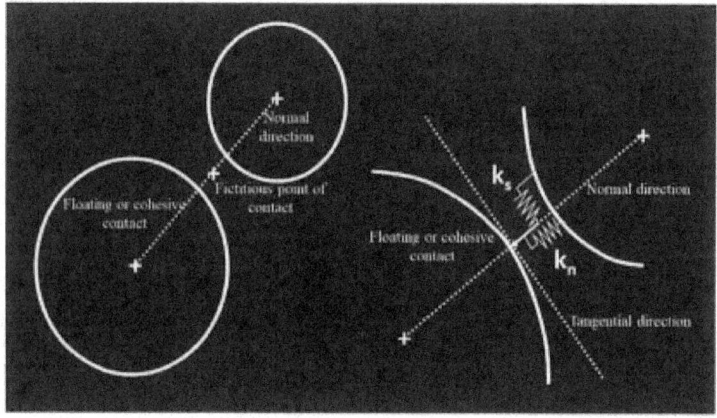

MOSTINI Planet Next Level

12. The fundamental Thermodynamic Sigma ZG Heavy Mechanical Pendulum and Oscillation Regulation

The oscillation of heavy pendulum interacts with The reactivities of fundamental thermodynamic on the quest to favor this reactivity to lowest Energy.

When the pendulum oscillates and reaches the equilibrium passage position, the speed is always maximum. This speed is maximum because of the optimal thermodynamic position, which is a stable position having the least energy of the pendulum's trajectory. The

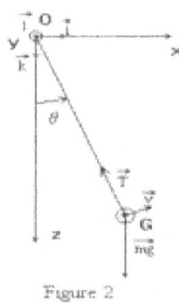

Figure 2

fundamental thermodynamic force governs the world.

MOSTINI Planet Next Level

Elastic interaction between two-point charges and thermodynamics. The thermodynamic electrostatic interaction between two point charges. Two very small identical spheres suspended by insulating wires are charged with the same amount of electricity. They repel each other by taking the position of thermodynamic equilibrium with the favorite position and less energy. The fundamental thermodynamics force governs the world.

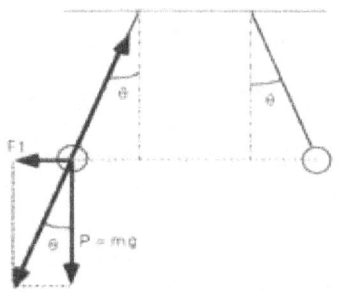

The fundamental thermodynamics force and Entropy (microcrack extension stress stability) are inversely proportional to two charges that repel each other until reaching the thermodynamic equilibrium, the position

MOSTINI Planet Next Level

with low energy. These small identical spheres exert a repulsive force on each other if the system's energy increases as they get closer. Otherwise, the force is attractive. Because the thermodynamic force rules the world, and the fundamental thermodynamic force and microcrack extension stress stability are inversely proportional and interact.

13. Osmosis and Thermodynamic Regulation

In water, the dissolved molecules are distributed evenly. Solutes are distributed according to the concentration gradient. That is, they move from the most concentrated to the least concentrated medium. This phenomenon is called diffusion. The fundamental thermodynamic force governs the world. The solute moves from the most concentrated medium to fulfill the thermodynamic needs of the favor reaction. In the living system, the plasma membrane is a dynamic component of the cell, regulating the passage of the substance, the movement of a molecule of any substance from an area of higher to lower concentration(with or down its concentration gradient) until the equilibrium of the fundamental thermodynamic force is reached.This diffusion process is also known as passive transport because the cell does not have to expend any energy (Entropy) to occur.Therefore, the fundamental thermodynamic force is performed in this process, which is the equilibrium state of low energy. Moreover, the fundamental thermodynamic

MOSTINI Planet Next Level

process favors chemical reactions

In case of several phases to reach an equilibrium state. Obviously, the chemical system in several phases can coexist. Each constituent will transfer from the phase in which its chemical potential is highest to the phase in which its chemical potential is the weakest until these chemical potentials equalize. Coulomb's law looks at the force created between two charged objects. As distance increases, the force and electric fields decrease. This simple idea was converted into s relatively formula. The force between the objects can be positive or negative, depending on whether they are attracted to each other or repelled. Obviously, in the infinitely large and macroscopic scale, entities with mass Ml and M2 attracts, when the energy between two mass entities increases, they repel with reactivities of Geo-thermodynamic Sigma ZG, the fundamental centripetal force and favor energy between Ml and M2, the isolated mass Ml and M2 increase energy in their surroundings, then they attract to reach the lowest energy. The fundamental Thermodynamic centripetal force governs the universe. This reactivity justifies the rotation of

MOSTINI Planet Next Level

planets around the Sun, the rotational reactivities of celestial entities, namely: Galaxies, Galaxies clusters, Solar systems.

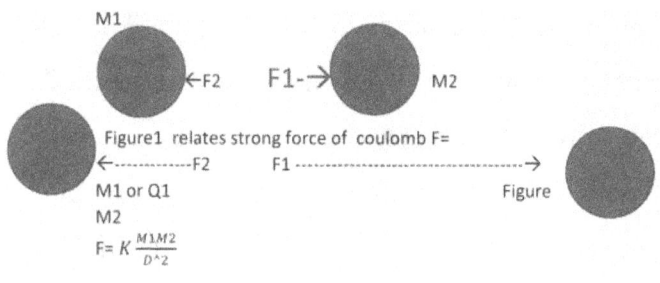

Figure1 relates strong force of coulomb $F= K\frac{M1.M2}{D^2}$

$$F= K\frac{\Sigma\Sigma Q1Q2}{D^2}$$

F1 is the thermodynamic centrifugal force of M1 exerts on M2 to reach lowest energy

F2 is the thermodynamic centrifugal force of M2 exerts on M1 to reach lowest energy

Mass = (Σ Heat + Σ Light + Σ Radiant + Σ electromagnetic + Σ nuclear + Σ sound) energies.

(M1 or M2 or mass energies)

M1 and M2 repel each other because their stored

MOSTINI Planet Next Level

energy or mass energy is high, so they repel to reach lowest energy. Afterwards, they attract because their stored energy perturbate their surrounding (Geo-thermodynamic). Also, the gravitational centripetal exerts between M1 and M2 are inversely proportional and relates their distance mass placement to lowest possible energy

M1 exerts on M2 gravitational centripetal force

M2 exerts on M1 gravitational centripetal force

The movement of repulsion and attraction between M1 and M2 creates the rotational revolution in theirs orbital.

F is attractive centripetal gravitational force between Ml and M2

Ml is the mass of first entities

M2 is the mass of the second entities.

The mass Ml and M2 are proportional to gravitational centripetal force entities. They are proportional to the heat produced, proportional to radiant energy, proportional to light energy produced, proportional

MOSTINI Planet Next Level

to electromagnetic force that induces the potential energy. Proportional to sound produced. The Gravitational centripetal force of Ml and M2 is inversely proportional to fundamental thermodynamic centrifugal force.

Mass = stored energy or potential energy

Mass = (Σ Heat + Σ Light + Σ Radiant + Σ electromagnetic + Σ nuclear + Σ sound) energies.

K = Gravitational constant between Ml and M2

Gl is the gravitational center of Ml

G2 is the gravitational center of M2

K is the reactivities of Gravitational force constant between M1 and M2.

$$K = \frac{G1+G2}{2}$$

The gravitational centripetal force between Ml and M2

$$F1 = F2 = K \frac{M1 M2}{D^2}$$

MOSTINI Planet Next Level

MOSTINI Planet Next Level

Interpretation Thermodynamics

Positive charge and negative charge attract each other and move towards each other. Similar charges such as two positive charges push each other. You also need to understand that the force between objects becomes stronger and weaker as they move apart. The fundamental thermodynamic force Centrifugal favors all reactions and energy. You could yell at someone from far away, and they would barely hear you. If you yelled the same amount when you were together, it would be more powerful.

14. Coulomb's Work by the Principle Of Regulation

Charles Augustin de Coulomb was a French scientist working in the late 1700' s. A little earlier, a British scientist named Henry Cavendish came up with similar ideas. Coulomb received most of the credit for the work because Cavendish did not publish all his work. The fundamental Thermodynamic force governs the universe.

MOSTINI Planet Next Level

The fundamental thermodynamic force and Entropy are inversely proportional. When you have two charged particles, an electric force is created. If you have a larger charge, the force will be charge be larger. If you use those two ideas and add the fact that charges can attract and rappel each other, you will understand Coulomb's Law. It's a formula that measures the electrical force between two objects.

$$F = \left(K \frac{\Sigma Q1 \Sigma Q2}{r^{\wedge}2} \right)$$

F is the resulting force between two charges.

The distance between two charges is R

QI and Q2 are values for the amount of charge in each of the particles. Scientists use Coulomb as units to measure charge. K is coulomb constant: 9. IO^ 9 Nm^2/C^2 The particle placed in electrical fields is moving toward decreasing potential. The fundamental thermodynamic force of all reactions, Chemical, Bio-chemical, and mechanical.

MOSTINI Planet Next Level

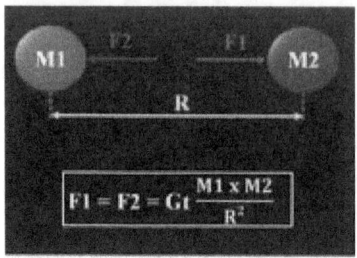

$$F = \left(K \frac{\sum Q1 \sum Q2}{r^{\wedge}2} \right)$$

Fl=F2 is the resulting force between stars and planets, and Stars, K is the constant thermodynamic

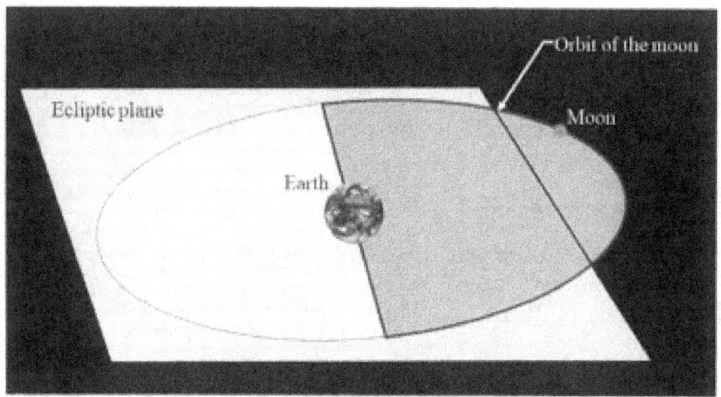

inversely proportional to the mass, QI planet charge, Q2

MOSTINI Planet Next Level

15. The Fundamental Thermodynamic Force Regulation and Helium Balloon

A helium balloon rises in the Earth's atmosphere because helium has a lower density than air. To put it another way: a volume of air weighs more than the same volume of helium. The balloon rises because all the heavier air is pulled down below it. This is a Buoyancy thermodynamic capacity

The balloon rises because of gravity fact. The gravity or the fundamental Thermodynamic centrifugal force pulls up the balloon directly through the atmosphere in favor and thermodynamic equilibrium and state of low energy. The thermodynamic law favors this buoyancy in relation with currents, Coriolis or gravity thermodynamic gradient. Also, in the deep land, volcanic lava flows, volcanic flames discharges, the rise of magmas, and the eruption of materials (gas and lava) from this magma on the surface of the Earth's crust or other planets, Stars or celestial entities are purely gravitational and

MOSTINI Planet Next Level

thermodynamic activities process to regulate and reduce energy and temperatures Obviously, from the steno sphere (1200 degrees), the rigid envelope of the terrestrial globe. We have some thrusters' firings of volcano the fundamental thermodynamic force and Gravity is the state of low energy; therefore, the gravitational traction could happen in any favor direction. The gravity is not always linked to elements' mass attraction but in the thermodynamic law and equilibrium and favor reaction. The low state of energy. The fundamental thermodynamic force is gravitational force activities, which favor all reactions, interactions in the universe. Thermodynamics and fluid mechanics.

During scuba diving, it is found that the pressure of the water increases with the depth. The water pressure exerted on a submarine at the bottom of the ocean is considerable. Similarly, the water pressure at the bottom of a dam is much greater than near the surface. Also, the universe's fundamental thermodynamic force directs and always acts to bring any object or plunger back to the surface. Thermodynamics is a universal force that reduces any physical, chemical or biochemical, mechanical interaction to a very low energy state or favors chemical, biochemical, and mechanical reaction. Anybody plunges into a fluid receives from this fluid a fundamental thermodynamic force (thrust) vertical upwards whose intensity is equal to the weight of the volume of this displaced fluid. (This volume is, therefore, equal to the immersed volume of the body or object.) This fundamental thermodynamic force also acts on fluid statics or hydrostatic, fluid dynamics, hydrodynamics, aerodynamics, aeronautics, meteorology, climatology, and aerography.

16. The Fundamental Thermodynamic Force And Buoyancy

Archimedean thrust is the fundamental thermodynamic universal force that undergoes a body immersed in whole or in part in a fluid (liquid or gas) subjected to the fundamental thermodynamic and gravitational field. This gravity is also the thermodynamic force that tends to bring back the object towards the surface for a state of very low energy exerted on the body. This

MOSTINI Planet Next Level

force comes from the increase of the pressure of the fluid with the depth: the pressure being stronger on the lower part of an immersed object than on its upper part, it results from a push which is the globally vertical action of the thermodynamic fundamental upward to accommodate or reduce to a low energy level the forces exerted on the body. It is from this thrust of this pushing thermodynamic force that the buoyancy of a body is defined. In a liquid, the bodies are subjected to the fundamental force thermodynamic force (Archimedean thrust). The bodies have different buoyancy depending on their density.

Buoyancy is the fundamental thermodynamics force that exerts a vertical thrust, directed upwards, that a fluid exerts on an immersed object. The fluid can be a gas as well as a liquid. Buoyancy obeys the law of gravitational thermodynamics that tends to orientate or direct the material and immaterial objects in a state of low energy in harmony with gravity as a state of low energy. According to the ratio between the real weight (Pr) and the Archimedean thrust (Pa), we distinguish positive buoyancy body: the object rises (Pr <Pa) negative buoyancy body: the

object is flowing (Pr> Pa) zero buoyancy body: the object floats between two waters. (Pr = Pa)

17. The Thermodynamic Properties and Regulation Between Earth and Moon

As you climb up to the top of a mountain, the air contains less and less oxygen. This is because the air pressure is decreasing.

Up to about 80 or 85 km altitude (hence beyond the stratosphere), the proportions of the various components of the atmosphere (nitrogen, carbon dioxide, oxygen, etc.) do not change, or little (air still contains 21% oxygen and 78% nitrogen, for example), because the air is well mixed. When the pressure is lower, the same volume of air contains fewer molecules, less oxygen, and less nitrogen and less carbon dioxide. Thus, on Mount Everest, the highest peak of the Earth, which rises to 8,848 m, the atmospheric pressure is one-third of the sea level. The same air volume (for example, each breath of a mountain)

MOSTINI Planet Next Level

contains three times fewer molecules.

Above 85 km altitude, these proportions change because of the components' molecular weight: the heavier are becoming rarer faster than the light ones. But even planes do not fly so high. Naturally, on the moon, the moon's atmosphere is extremely tenuous and even insignificant compared to that of the Earth. Its density does not exceed one hundred millionth of that of the Earth, at sea level. In practice, the Moon can almost always be considered as surrounded by the void. We notice that the entropy of the earth is greater than that of the moon. The planet earth has a very large entropy for two reasons; its electrostatic potential because of its mass, which is greater than that of the moon. Its chemical potential due to the presence of the atmosphere (the exchanges of the chemical elements) (air still contains 21 % oxygen and 78% nitrogen). The Entropy of the system is always proportional to the mass. Obviously, the moon is empty while the planet earth has an atmosphere. We say the entropy of a system or an entity is proportional to its mass. In this case, the entropy is the set of the chemical potential and electrostatic

MOSTINI Planet Next Level

potential. The entropy of the moon is much inferior to that of the Earth because the moon is empty. It's entropy is close to the primordial universe before the beginning of all things.

Based on the laws characterizing the universe that we have detected with the galenic structure of the Antimicrobial cream, which states: "The fundamental thermodynamic force and the Entropy are inversely proportional."

$$\lim_{T \to 0} (\Delta S)_T = 0 \qquad \left(\frac{\partial E}{\partial S}\right)_{V, n_i} = 0$$

First, the fundamental thermodynamic level is very important and higher on the moon than on the earth. However, the entropy level is very small on the moon than on the earth. On the other hand, the Earth's entropy level is very important and higher on the Earth than on the moon. The entropy level of the moon is very small than on the Earth. However, the fundamental thermodynamic force is very large, important and higher on the moon than on the

MOSTINI Planet Next Level

Earth. The moon is more thermodynamically stable than the Earth.

1. The Fundamental thermodynamic force on the moon is very large or very important than on the Earth.
2. Before the beginning of the Universe or before the creation, the thermodynamic fundamentals force was very stable to a high level and maximal. The universe was pure and empty

$$\left(\frac{\partial E}{\partial S}\right)_{V,n,} = 0 \qquad \lim_{T \to 0} (\Delta S)_T = 0$$

The entropy of the universe was zero at the beginning, the universe was like an era insulator (the third thermodynamic law). Obviously, we have:

1. The entropy of the Earth is a very important level and a large amount. However, its fundamental thermodynamic force is small.
2. The entropy of the moon is small. However, its

MOSTINI Planet Next Level

fundamental thermodynamic force is very stable and large, and the moon has a great and strong thermodynamic force.

3. The moon's Fundamental thermodynamic force is very large or very active and important than on the Earth.
4. Before the beginning or the creation of the Universe, the thermodynamic fundamentals force was very stable, important, large and maximal. The universe was pure and empty.

On the other hand, the Entropy of the universe was zero at the beginning as an insulator (the third thermodynamic law). The fundamental thermodynamic force and

Entropy and microcrack extension stress stability are inversely proportional.

$$\lim_{T \to 0} (\Delta S)_T = 0 \qquad \left(\frac{\partial E}{\partial S} \right)_{V, n_i} = 0$$

MOSTINI Planet Next Level

MOSTINI Planet Next Level

18. The Astronaut Thermodynamic Properties In The Space

When they are in their shuttle, the astronauts give the impression to be light and to float: they are weightless. Obviously, the fundamental thermodynamic force is very high level, stable, and very active in space. The Fundamental thermodynamic forces orients runs and always directs all universe entities (physical, chemical, biological, humans) in the physical and electromagnetic state level of low energy towards a low energy state. In the space, all entities, bodies, astronauts float. Actually, on the planet Earth, there is a force that attracts us like a magnet. On the ground, the fundamental thermodynamics force is weak than on space or the moon. And people do not float. This is the effect of the fundamental gravitational thermodynamic force always orienting the objects or entities (chemical, biological, biochemical, physiochemical) towards the state or low energy framework. This fundamental thermodynamic force keeps us on the ground: if we jump, we always fall to the ground.

MOSTINI Planet Next Level

When we touch the ground, our body is always moving towards a low energy state and very low energy: it is gravity linked to a thermodynamic movement of low energy. But if we go from the ground to space, we are in a state of weightlessness or the state of great, high, more thermodynamic stability force. The astronauts in their shuttle are therefore falling but around the Earth. And if you do not have the impression of falling by seeing them, it is because the shuttle, in which they are, falls at the same speed as them.

Obviously, on Earth, sometimes we are in a state of weightlessness and have the impression of no longer touching the ground: by car, for example, when we pass on bumps or in some rides, like the roller coaster, when we go down very quickly. In space, the thermodynamic fundamentals have a very high-level amount. The fundamental thermodynamic force exerts a superficial tension, the tension surface on the Astronauts, shuttle, all object and all celestial entities (Moon, Sun, clusters, Galaxies) But it is not always easy for astronauts to live in weightlessness: objects float everywhere, water curls up,

MOSTINI Planet Next Level

and they must cling to their beds to be able to sleep. They must also do a lot of sport because otherwise, their muscles may shrink if not using much of their legs. The fundamental thermodynamic force is always orienting people, objects and celestial entities in a state of low energy.

The thermodynamic fundamentals force can only orient an object or entities in all directions according to the low energy state, always leaving and orienting towards the chemical potential and the lowest electrostatic potential and lowest energy because of the kinetic stability. Also, considering the translational movements of the plane's physics, we have four fundamental force interactions that combine a center of gravity with thermodynamic equilibrium. On the one hand, we have the drag and weight force strictly controlled by the fundamental thermodynamic force, leading and resulting in low energy. We also have Entropic forces (aerodynamic lift or thrust and traction). The fundamental thermodynamic force and Entropy co-exert inversely proportional forces to maintain the aircraft in their thermodynamic equilibrium translation movements.

MOSTINI Planet Next Level

Obviously, the fundamental thermodynamic force (the drag and weight forces are inversely proportional with Entropy (aerodynamic lift or and traction)

MOSTINI Planet Next Level

19. The Properties Of The Fundamental Thermodynamic Force Of The Apple On The Moon And On The Earth

Question: Why the apple kept and released at a distance H from moon's ground does not fall when the same is placed at the same height H on Earth planet, The apple fall?

Response: The answer to this question allows us to talk about two reasons. The first is the influence of the fundamental thermodynamic centrifugal force and the Entropy centripetal force, and the second is the effect of buoyancy

As we previously explained, the amount of thermodynamic fundamental force on the moon is especially important and very important. On the other hand, the amount of his Entropy is very weak. The fundamental thermodynamic force and Entropy are inversely proportional. When the apple is placed at a certain height H of the lunar soil, the apple does not fall. The important amount of the fundamental thermodynamic force keeps by

MOSTINI Planet Next Level

exerting a huge superficial tension of the apple then the apple is suspended state with very low energy. This is also characterized by the absence of forces of friction forces on the moon. The apple is suspended by the presence of the high amount of fundamental thermodynamic force. Further, in space, the fundamental thermodynamic force keeps the sun and all celestials' entities suspended because the amount and level of the thermodynamic force in the sun is equal to the superficial tension of the sun or Sun tension surface. The sun and all the space entities are kept in a suspended position, which is stable and low energy. All spatial entities are kept suspended by the fundamental thermodynamic force while giving them a round shape. This round shape is made possible by superficial pressure or surface tension exerted by the surfaces' fundamental thermodynamics. Considering the apple, the sun or all the spatial entities. Besides, all the celestial entities are round because of this great thermodynamic force, which is very high, keeping them in a state of very weak and low energy. Also, the fundamental thermodynamics force exerts huge, high, and larger amount enough of surface tension or

MOSTINI Planet Next Level

tension surface on the moon to capture the apple, and the apple does not fall. This is the state of low energy. Moreover, when the same apple is placed on the planet earth at the same height and same condition, it falls because the fundamental thermodynamic force is weak than in the moon then brings it back to the ground with a low energy level.

When the fundamental thermodynamic force maintains suspended celestial entities, the apple will act on the surface tension. The force of the superficial tension thermodynamics exerts on the entity or the apple is equal to the force exerted by the interface of the apple or the Celestial entities, and they remain suspended.

MOSTINI Planet Next Level

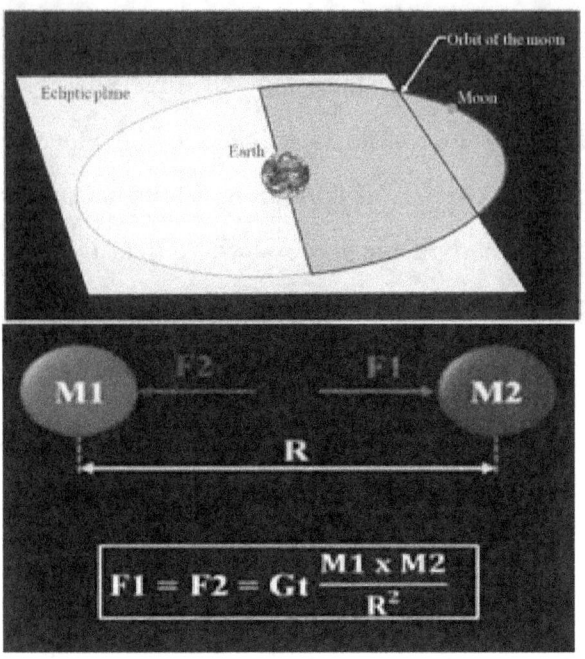

$$F1 = F2 = \left(K \frac{\Sigma Q1 \Sigma Q2}{r^{\wedge}2} \right)$$

The geo-thermodynamics or Potential Sigma ZG of a geographical environment(Potential sigma ZG) is a force of the fundamental thermodynamic of the universe which consists in maintaining a geographical place in a situation

MOSTINI Planet Next Level

of low possible energy by regulation instability(Radioactivity, magnetic moments, electric moments) to reach the level lowest stability of universe primordial

$$\left(\frac{\partial E}{\partial S}\right)_{V,n_i} = 0 \qquad \lim_{T \to 0}(\Delta S)_T = 0$$

Geo-thermodynamic or the potential sigma ZG always regulates celestial and universe entities to conform to the geographical and environmental area, co-interaction, energies, and reactivities in the strict condition of lowest energy. Obviously, the planetary or star electrostatic regulation levitation reverses magnetic poles regulation (the differential motion of the solid and liquid par of Earth' core), the Radioactivity regulation, electrical moments regulation, and magnetic moment regulation. In the geothermal referential, all planets as well as the planets Earth (Telluric planets), and all entities, for instance, Satellite (Moon) are considered as masses punctual

MOSTINI Planet Next Level

positively charged and positive gradient (The Core of the earth and the moon have a positive gradient metal, iron ions in curvilinear motions), which exert each other and electrostatic potential levitation force. Because of the electrostatic potential of each other (Earth and Moon), They repellent each other by taking the position of thermodynamic equilibrium with the favor position and less energy(orbited). In the space without friction, the resulting movement of this equilibrium is the ellipse that represents the lunar orbit, the fundamental thermodynamics force governs the world. The fundamental thermodynamics force and Entropy (microcrack extension stress stability) are inversely proportional. Because of the absence of friction, the energy induced by the fundamental thermodynamic force and microcrack extension minimum stress stability must be available to maintain the earth and the Moon in the thermodynamical interaction keeping them in the situation of low energy and they rotate with the constant speed. Obviously, two Celestial entities repellent each other until reaching the thermodynamic equilibrium, the position with low energy. The celestial entities exert a repulsive force on

MOSTINI Planet Next Level

each other if the energy of the system increases as they get closer. Otherwise, the force is attractive. When they reach the orbital minimum and stability level, they rotate under their centripetal mass kinetics attraction and centrifugal electrostatic repulsion effects. The co-action of centripetal and centrifugal movements generates electromagnetics kinetics thermodynamics moments in a stable orbital and rotational movement with circular, elliptical stability on balance harmonic movement lowest energy on the quest of search primordial universe realities with the lowest stability Because the fundamental thermodynamic force rules the world. Also, the fundamental thermodynamic force and microcrack extension stress stability (Entropy) are inversely proportional and co-interact mutually and continuously on the quest to reach the condition of low energy confirming the condition of the primordial universe with high stability.

MOSTINI Planet Next Level

20. The Thermodynamic properties between earth and moon

$$\lim_{T \to 0} (\Delta S)_T = 0 \qquad \left(\frac{\partial E}{\partial S}\right)_{V, n_i} = 0$$

Compared to the planet earth, for several reasons, the moon's surface is essentially solid compared to the Earth composed of water. The mass of the moon is small, so its Entropy is small compared to the entropy of the Earth. The Entropy is always proportional to the mass of the system. Therefore, the thermodynamic fundamental is very important and high on the moon than on the planet earth. The fundamental thermodynamic and Entropy are inversely proportional. The third principle of thermodynamic is associated with the descent to its fundamental quantum states of a system whose temperature approach a limit which

The circular motion of planets, sun, electron, neutron stars, galaxies, clusters, super clusters, solar systems, planets, stars, all entities obey the thermodynamic

MOSTINI Planet Next Level

law and the living system. The thermodynamic law is the fundamental force in the universe, which controls all entities as well as animal, plant, human, and their metabolism system.

21. Manifestation and Influence of Second Thermodynamic in The Living and Human Body (The second Thermodynamic Law)

The fundamental Thermodynamic law governs the universe. The common sense, energy refers to anything that makes it possible to do work, produce heat, create light, modify chemical species, and generate movement. Therefore, energy is a physical, numerical quantity, which has no general character because associated with a concrete situation. It is neither fluid nor a substance. We speak of movement energy, electrical energy, chemical energy, thermal energy, nuclear energy, hydraulic energy, wind power. When death occurs, all biochemical reactions in the body reach the strict stability level of chemical bio-

MOSTINI Planet Next Level

reactions. The energy of body system equals zero, the Physiological Entropy equals zero, Temperature zero. The thermodynamic in the human body reaches the strict lowest stability level.

$$\lim_{T \to 0} (\Delta S)_T = 0$$

We can explain this thermodynamic law's manifestation and influence in the human body and all living systems (animals, plants) by fatigue, hunger, sleeping, disease, aging, and finally death. The fundamental thermodynamic force gives the round shape of planets, stars, sun, and all celestial entities exert a superficial tension on the surface), the aging of material (the aging, and the fading of the wall of the houses, the buildings, the rust of iron materials, the deterioration of other metals). Obviously, a mature male animal or a mature human who abstains from sexual activity for a long time fulfills the bio-thermodynamic need bringing biochemical reaction by the sexual drive activities. The sexual drive is justified by nocturnal ejaculation during sleep or spermatozoa's

MOSTINI Planet Next Level

recurrent presence in the urine. In this situation, the individual is complying with this thermodynamic need of the living body. The thermodynamic law complying all living systems' reproduction process controls chemical and biochemical activities in the universe. Also, the evolution process driving by natural selection.

Natural selection is the mechanism by which species have evolved by adapting to their environment. This evolution and adaptation to the environment is work and quality made possible by the fundamental thermodynamics forces in the life of the species acting on the geographical environment and the physiology of the individuals to facilitate the daily activities and to improve the physiology of the species for a better adaptation in a sophisticated framework and to spend less energy both in the geographical environment and for the physiology of the species. The fundamental thermodynamic universal force always favors all reactions, interactions and activities in the universe and always tends to organize, rehabilitate all states of physical, chemical, biochemical reactions and activities with the lowest energy. Fundamental thermodynamic force

MOSTINI Planet Next Level

controls and determines the physiological, organoleptic, and physical condition such as the skin color, hair quality and size, weight, height, eyes color, race, the shape, and the nature of all species (animal and plant in all living system.

The fundamental Thermodynamic Force centrifugal governs the World. The fundamental Thermodynamic force and Entropy are inversely Proportional. This diagram shows the coordinated activity of the fundamental thermodynamics force and microcrack extension stress stability (Entropy) in continual, perpetual correlation , and these eco-activities of the fundamental thermodynamic force and Entropy governs the world, all the activities of the universe (mechanical movements of all, the living beings, any material object perceptible and not perceptible, existential obey to this law.

MOSTINI Planet Next Level

The fundamental Thermodynamics governs the world. We may practically display the correlation between the fundamental thermodynamic and macrocrack extension

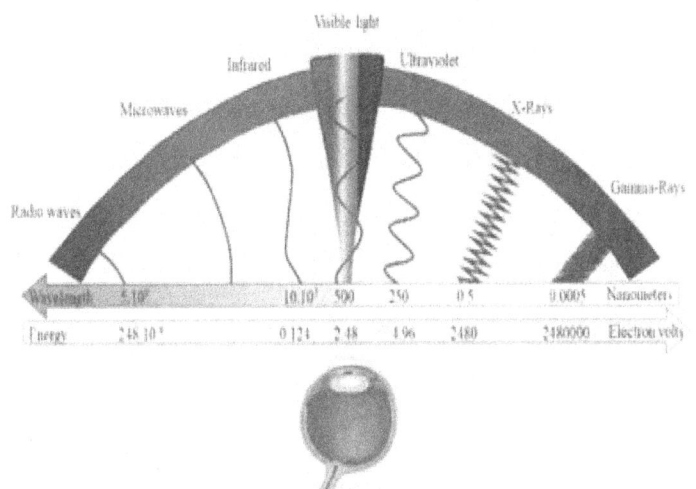

stress stability, which are inversely proportional. This perpetual correlation may be called the Quantum Metrix Antigravity.

22.The properties of the fundamental thermodynamic force and IMC or BMI body mass index(2nd thermodynamic law regulation)

MOSTINI Planet Next Level

In the human body, all biochemicals reactions tend to be less stable to more stable with increased time (aging) or when people are sick or tired. Obviously, all pathology in the human body justifies the states of low energy. Finally, death is a loss of complexity, a loss of organization. The chemical reaction reaches a stable level. A great loss of Entropy and disorder. However, the antimicrobial skin and its organic silicon challenge thermodynamic stability and creating physiological Entropy to heal skin pathology. The low stability of the skin in the human body may be: skin disease, wound, burn, leprosy, psoriasis, diabetic wound, skin cancer, impulse disorder control. According to galenic criteria, the healing model of antimicrobial is particular. It heals several diseases. In the universe, the Entropy is proportional to the masse

Thus, the body mass index, or BMI, makes it possible to know if the measured person has a normal corpulence, or if it is overweight, in obesity, thinness or undernutrition. Therefore, corpulence is very important to identify or predict the quality of health. In the human body,

MOSTINI Planet Next Level

the thermodynamic favor all biochemical reactions. The BMI may give medical information about the impact of thermodynamics in the patient's life to prevent disease. The BMI is very popular with health professionals because it allows them to obtain an objective result, insofar as it considers the person's size. This index is very much relative to the weight of the person, plus the weight is high, the thermodynamic body stability is effective with, so problems of diseases. The main purpose of BMI is to assess the health risks associated with weight. The normal scale and goal are to stay between 18.5 and 25. This is the normal scale of physiological Entropy, in the correlation of the fundamental thermodynamic force's good level. We have the ability to identify and quantify the thermodynamic level to prevent disease.

Indeed, the human body's fundamental thermodynamic force ability is the state of low energy between chemical and biochemical reactions. Therefore, all pathology represents the thermodynamic ability and impact in the human body, overweight, and obesity, particularly in terms of cardiovascular diseases, Blood pressure, diabetes,

MOSTINI Planet Next Level

sleep apnea and other metabolic diseases are reduced when a doctor is aware of patient BMI by advertising to challenge the thermodynamic stability and exercising to increase the physiological Entropy. The fundamental thermodynamic force and microcrack extension stress stability are inversely proportional. Also, the thermodynamic influences and impacts people when the BMI is less than or equal to 25. Conversely, a state of emaciation or malnutrition can also lead to health problems. Thermodynamics influences the human body. The fundamental thermodynamic force impacted and influenced Mankind during the human's evolution period. Moreover, aging, sleeping, fatigue, impulse disorder, disease, Alzheimer, all kind of pathology is the influence and impact of thermodynamic law stability in the body. Diseases like Diabetes, tension, aging gravitational or thermodynamic diseases are not hereditary or genetic transmission but are transmitted from generation to generation due to a bad custom and dietetic habits culture. The proportion of silicon always decreases with age, and most living beings are born with about 0.7g to 0.8g of

MOSTINI Planet Next Level

silicic acid or Silicon. The Silica and ortho-silicic acid are decreasing as we grow older and finish more and more are older. Because the Ortho-silicic acid and silicon challenge the thermodynamic stability in the human body (lowest energy of biochemical reaction), the risk to develop the incurable disease is higher with aging (elder). The living beings develop the risk and are victims of gravitational and thermodynamic diseases (Diabetes, blood pressure, Alzheimer's, Osteoporosis, impulse disorder control) generated by diminution and decreasing of Silica and ortho-silicic in the human body. The fundamentals thermodynamic force and microcrack extension stress co-interact stress to ensure body organs' physiology and function, body organs' physiology and function, including multiplication, DNA duplication, respiration, and harmonious development of the body (living system).

When you inhale or breathe, the diaphram and muscles of your ribs contract, and create negative pressure or inside your chest cavity. The negative pressure (fundamental Thermodynamic activities) draws the air that you breathe into the lungs.

MOSTINI Planet Next Level

clxxx

MOSTINI Planet Next Level

23. Thermodynamic Law Mineral Regulation Mineral Through Organic (Ortho-silicic acid In The Formation of Living Cell, DNA, and Life the Genetic Takeover)

Genetic takeover is about the origins of life on earth. Obviously, in the earth early time. The fundamental force thermodynamic law in the control of silicon, ortho-silicic and clay reactivities interacted with the entropy(microcrack stress stability) led the chemical evolution, in which a 'primeval soup' leads to the formation of prebiotic elementary.

Obviously, the primeval soup was clay(in composition with inorganic colloids, silicon, ortho-silicic acid, and mineral element; (Na^+, Ca^{2+}, Mg^{2+}, Zn^{2+}, Al^{3+}, Fe^{3+}, Fe^{2+}, Cu^+, Cu^{2+}, Ti^{4+}) in seawater (in composition with sodium chloride, mineral elements The ortho-silicic acid and silicon organic of clay mineral; (Na^+, Ca^{2+}, K^+, Mg^{2+}, Zn^{2+}, Al^{3+}, Fe^{3+}, Fe^{2+}, Cu^+, Cu^{2+}, Ti^{4+}). The ortho-silicic acid and silicon organic of clay mineral, simulated

MOSTINI Planet Next Level

ancient seawater, clay, silicon organic and ortho-silicic forms a hydrogel, which performed Oxydoreduction of mineral elements most (iron, copper) with Hysteresis and stepwise activities creating a mass of microscopic spaces capable of soaking up liquids like a sponge. Over billions of years, chemicals confined in those spaces could have carried out the complex reactions, the supra-molecular assemblies, addition, transformation, and resonances (complex mesomeres or resonance reactions between cations mineral and colloids or negative complex) that giving many pre-biotic compounds and formed afterward, proteins, DNA, and probiotics in the presence of light and Sun energies, and eventually all the machinery that makes a living cell work. Clay that contained ortho-silicic acid and silicon organic hydrogels could have confined and protected those chemical processes until the membrane that surrounds living cells developed.

S_{sys} entropy of system > 0 implies that the system becomes *more disordered* during the reaction.

S_{sys} entropy of system < 0 implies that the system becomes *less disordered* during the reaction.

MOSTINI Planet Next Level

The Kaolin clay formula is:

$(Ca,Na,H)(Al,Mg,Fe,Zn)_2(Si, Al)_4O_{10}(OH)_2*H_2O$

The third principle of thermodynamic is associated with the descent to its fundamental quantum states of a system whose temperature approach a limit which is the creation of life is the result of thermodynamic regulation from minerals through the organic system. The passage from less energy to most less energy to conform reactivities in the condition of universe primordial

23. Electricity regulation activities in the living system(Animals and human)

Electricity is present in all the infinite universe and antimicrobial cream and living organisms (humans, animals, plants). Therefore, In the human body, at the microscopic level, the cells that make up the human body are bathed in a liquid containing all the nutrients, minerals and all elements necessary for their proper functioning. The cells' interior is separated from this extracellular medium by a lipid membrane called a plasma membrane, which

MOSTINI Planet Next Level

forms a tight barrier between the intracellular and extracellular compartments. Electrically charged atoms, called ions, can enter or leave the cell via transporters and membrane channels. Like an electric battery, it is the displacement of these ions between the inside and the outside of the cell and the resulting electronic imbalance at the origin of the cells' electrical activity. This electrical activity is important for neurons, of course, because it participates in the coding and transfer of information and other types of cells in the body, for example, by contributing to cellular metabolism. The Antimicrobial skin cream, the human body, and the universe present the same characteristic and properties. Therefore, Ortho-silicic and Silicon are present in the universe (Sun, Galaxies, Plants), the Antimicrobial skin cream, and in the living organs system (humans, animal, and Plant). They run electricity and control chemicals reactions. The clay Kaolin containing Silicon and Living organ are the products of the universe. They are representing the creation of the cosmos. The living organ as the Silicon and ortho-silicic acid and clay (recover the planet's surface) are the key to the

MOSTINI Planet Next Level

functioning of the universe.

24. Thermodynamic gas parameters regulation

Guy lussac 's law states that the pressure of the given mass of gas varies directly with the absolute temperature of the gas.

Guy-Lussacs' Law

- Guy-Lussacs' Law. The number of particles of the substance are held constant.

- $P_1/T_1 = P_2/T_2$

MOSTINI Planet Next Level

Before creation temperature and pressure, Helmotz (T,P) were zero.

How does Boyle's Law relate to breathing?

The Mechanics of Human Breathing. The relationship between gas pressure and volume helps to explain the mechanics of breathing. Boyle's Law is the gas law that states that pressure and volume are inversely related in a closed space. As the volume decreases, pressure increases and vice versa.

MOSTINI Planet Next Level

Charles' law describes the relationship between temperature and volume at a constant pressure. The ability to visualize individual gas particles' behavior in an enclosed space helps in understanding the mechanism underlying Charles' Law.

$V \propto T$ or,

$\dfrac{V}{T}$ = constant = K

Here K is a constant that deoends on the pressure of gas, the amount of gas and also unit of volume.

If V_1 and T_1 are the initial values of volume and temperature of a gas then,

$\dfrac{V_1}{T_1} = K$

Also, if the temperature is now changed to T_2 such that the volume changes to V_2

we can write,

$\dfrac{V_2}{T_2} = K$

or $\dfrac{V_1}{T_1} = \dfrac{V_2}{T_2}$ or $\boxed{V_1 T_2 = V_2 T_1}$

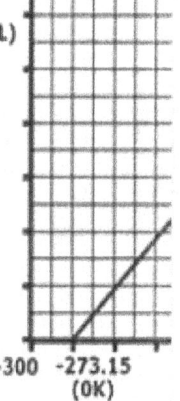

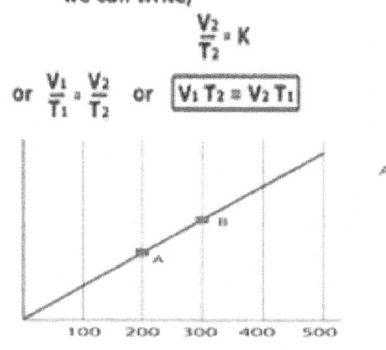

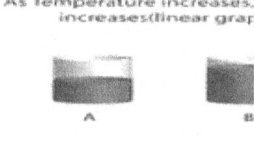

As Temperature increases, increases(linear graf

MOSTINI Planet Next Level

25. The physicochemical potential Sigma ZG Energy Barrier Zeta

A/ Potential Energy Barrier Sigma (ZG) or Thermodynamic Sigma (Z= 0) Energy

The universal fundamental thermodynamic force sometimes gives rise to kinetic stability (due to an energy barrier that is too high to overcome). To take the example of the diamond, even if the fundamental thermodynamics law force favors the transformation into graphite, the energy required to achieve it is so high that we observe the kinetic stability. In thermodynamics, a thermodynamic system is in thermodynamic equilibrium when it is at the same time in thermal, mechanical and chemical equilibrium. The local state of a thermodynamic equilibrium system is determined by the values of its intensive parameters, such as pressure or temperature.

More specifically, thermodynamic equilibrium is characterized by the minimum of a thermodynamic potential, such as Helmholtz free energy for constant temperature and volume systems or Gibbs free enthalpy for

MOSTINI Planet Next Level

constant pressure and temperature systems.

The process leading to thermodynamic equilibrium is called thermalization.

MOSTINI Planet Next Level

Chapter V
The Pollution of the Universe

1. Regulation of Universe Pollution Charge

The particle by the fundamental Thermodynamic Force. The third principle of thermodynamics is associated with the descent to its fundamental quantum state of a system whose temperature approaches a limit, which

$$\lim_{T \to 0} (\Delta S)_T = 0$$

defines the notion of Entropy.

When an electron is hit by a photon of light, it absorbs the quanta of energy that the photon was carrying and moves forward to a higher energy state. The electron

MOSTINI Planet Next Level

can then gain the energy it needs by absorbing light. If the electron jumps from the second energy down to the first energy level, it must give off some energy by emitting light. Moreover, the absorption and emission process in terms of an electronic structure according to the Bohr model, an electron absorbs energy under photons (light) and reaches higher levels. When the electron returns to lower energy levels, they release extra energy, and that can be in the form of light causing the emission of light. Obviously, an electron is in a less stable and more thermodynamically stable. The fundamental thermodynamic force governs the universe, universe, one time, universe. Moreover, the fundamental thermodynamic force and Entropy are inversely proportional.

2. Geo-Thermodynamic Lightning Phenomena on the Effects of Annihilation and Regulation

When separating positive and negative charges, a cloud gives rise to lightning. The fundamental geo-thermodynamic centrifugal force and the macrocrack

MOSTINI Planet Next Level

extension stress stability centripetal force always regulates the electric charges or the electromagnetic energy. When there is the disturbance of the charges. To regulate the geographical environments in the conditions of the primordial universe with zero energy. T = O, S = O and Zeta=O

$$\lim_{T \to 0} (\Delta S)_T = 0$$

The turbulence of a storm with its upward and downward currents represents the ideal environment for separating electrical charges. The fundamental geothermodynamic force and macrocrack extension stress stability favors the location of negative charge concentrations at the base of the clouds, while the positive charges begin in the upper parts

This phenomenon allows the electric field to form and fill up between the cloud and the soil by the principle of geo-thermodynamic Sigma ZG and microcrack extension stress stability of the solar system (the sun and planet relationship in the levitation, interaction, and

MOSTINI Planet Next Level

attraction reactions).

Similar charges repel each other, and the opposing charges attract each other, the negative charges begin to spread near the cloud base, much like a shadow. Positive charges tend to focus on objects and heights. A cloud-ground eclair begins when a negative charge makes its way to the ground. This phenomenon tracer bond can go up to 100m each. This leap tracer can go in several directions under the thermodynamic stability control. In response to the negative charge from the base of a cloud, positive charge currents begin to move upward from the ground. This is an ascending tracer. The fundamental geo-thermodynamic force and macrocrack extension stress stability always regulates the positive and negative charges to reach the conditions of the primordial universe. $T = OK$, $S = 0$ (Entropy), $Z = 0$ (zeta)

MOSTINI Planet Next Level

3. Thermodynamic universe regulation

$$\lim_{T \to 0} (\Delta S)_T = 0$$

The fundamental Thermodynamic governs universe and universe. The fundamental Thermodynamic force and Entropy are inversely proportional.

4. Geo-thermodynamic or Potential Sigma ZG Regulation and Kinetic Moment

When the fundamental thermodynamic force regulates a reaction, another problem arises which results from the consequence of the first problem solved. The fundamental thermodynamic force of the universe exerts a negative pressure in all the universal reactions and universal entities, to bring them back to the physical and chemical conditions of the primordial universe.

MOSTINI Planet Next Level

$$\lim_{T \to 0} (\Delta S)_T = 0 \quad \left(\frac{\partial E}{\partial S}\right)_{V, n_i} = 0$$

Because the fundamental thermodynamic and macrocrack stress stability are inversely proportional, the kinetic moment, gravitational moment, electric moment, and magnetic moments are induced, modeled, and initiated for the search of physical and chemical conditions for the primordial universe. We have T=0K, the Earth remains in orbit around the Sun. It has kept the rotational movement of the molecular cloud that formed the solar system. It is thus in orbit around our star.

Then our planet turns on itself also because it has kept the rotational movement of the protoplanetary disk and the reactivity of the potential sigma regulation introduced by the principle of the continual interaction between the fundamental thermodynamic force and the macrocrack extension stress stability on the quest to reach the lowest energy with condition universe primitive. The planets in orbit and the electromagnetic force playing the role of

MOSTINI Planet Next Level

emergent gravity under the action of mutual interaction and antagonist of the fundamental thermodynamic force and macrocrack extension stress stability. Here, the planets would be kept in circular orbit with the lowest energy stability of reactivity and action of the balance between centrifugal force, which tends to keep them away from the sun, and centripetal force generating electromagnetic force, which attracts them to the sun and kept the orbital movement interaction between Sun and the Earth. The universal regulation may be the magnetism, radioactivity, reversing Magnetic pole, electric moments, magnetic moments

5. Potential Sigma ZG and Magnetism Regulation

The magnetic field is the fundamental thermodynamic force regulation exerted between the positive and negative electric instability charges to obtain a zero-energy value as the values of the primordial universe.

MOSTINI Planet Next Level

$$\left(\frac{\partial E}{\partial S}\right)_{V,n_i} = 0$$

Magnetism arises from the thermodynamic instability between the negative electric charge currents and the positive charge currents. This force tends to reduce the charges of negative and positive electrical current in a state of thermodynamic and geothermodynamic stability (The Potential Sigma ZG) under the conditions of the primordial universe (as an insulating era). T=0k, S=0 (Entropy), Z=0 (Zeta).

Obviously, the Earth revolves on two magnetics electricity magnetics differential motion moments. The positive liquid core part positive current magnetism differential motion moments caused by the differential rotational motions moments (iron ions), by regulation principle, when the positive magnetism is action, the negative magnetism arise because they co-interact. Therefore, the Earth planet arise the colloidal magnetic, which is the negative magnetic with the opposite sense with the positive magnetism.

MOSTINI Planet Next Level

MOSTINI Planet Next Level

6. Potential Sigma and Radioactivity Regulation

The radioactivity field is a fundamental thermodynamic force regulation exerted in geographical areas and all entities material, physical living system represent universe perturbation and instability of the positive and the negative electric particles and charges to obtain a zero energy value as the values of the primordial universe.

$$\left(\frac{\partial E}{\partial S}\right)_{V,n_i} = 0$$

All living system in the world or Entropic systems (creatures process system or all living system) generate radio-activity field in the microscopic field because of the fundamental thermodynamic or the geo-thermodynamic regulation exerted between positive, bio-positive and negative, bio-negative to obtain zero energy value of the primordial universe.

MOSTINI Planet Next Level

All entities generate radioactivity by geo-thermodynamic regulation in the microscopic scale and macroscopic scale. The fundamental thermodynamic force governs universe, universe one time.

7. Potential Sigma ZG Regulation and the Reversing Magnetic Earth

Earth's magnetic changes or reverse result from the interaction of the thermodynamic force and micro-crack extension stress stability, at what is called the "Earth dynamo." This is caused by the differential motion of the solid and liquid parts of the Earth's core. Over time because of kinetics regulation for the lowest energy stability, must alternate to renew the moments and rotational magnetic confinement movements of the Earth, make them and make them much more dynamic by creating more stable energy. The states requiring less energy so that the moments of the rotations are always renewed and constant in over 750,000 years.

For instance, a pendulum weighing at the passage of

MOSTINI Planet Next Level

the equilibrium position which is the most stable the pendulum rotation speed is always maximum by the effect of the fundamental thermodynamic force and microcrack extension stress stability this equilibrium position is conserving and keeping rotational pendulum movement. Moreover, the force between two wires, each of which carries a current, can be understood from the interaction of one of the currents with the magnetic field produced by the other current.

For example, the force between two parallel wires carrying currents in the same direction is attractive. It is repulsive if the currents are in opposite directions. Two circular current loops located one above the other and with their planes parallel, will attract if the currents are in the same directions and will repel if the currents are in opposite directions. The magnetism field positive have opposite direction and sense with magnetism negative. Because the Earth is composed of two magnetism in the opposite direction, the reversing magnetic Earth is the equilibrium position in the scale 750,000 to 800,000 years. When positive magnetic field is action, the negative magnetic

MOSTINI Planet Next Level

field is also generated in the opposite sense. They exhibits the instability to which the fundamental thermodynamic force and microcrack extension stress stability must regulate to interfere in the condition of low energy, zeta zero energy in the condition of the primordial universe.

The figure below represents an elementary principle.

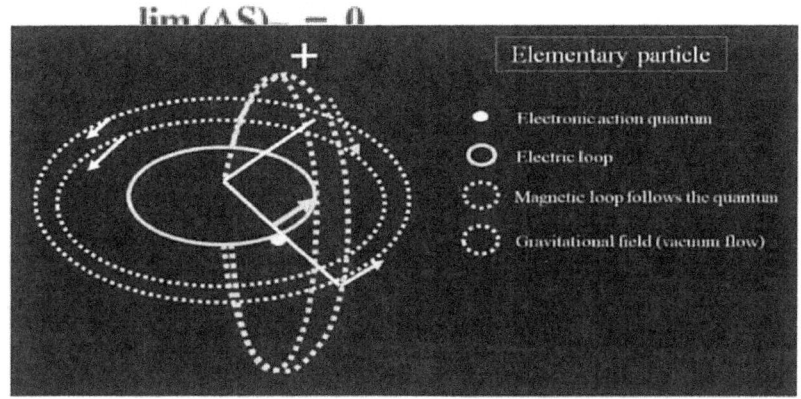

The fundamental Thermodynamic Force governs universe and the fundament thermodynamic and Entropy are inversely proportional. This seemingly complicated force between current loops can be understood more simply by treating the fields as though they originated from

MOSTINI Planet Next Level

magnetic dipoles. There is an interaction of the fundamental thermodynamic force and macrocrack extension stress stability that occurs between the magnetic loops of cations precisely the iron at the nucleus and the magnetic loops precisely anions of the colloids and Silicon from the mantle and surface.

The cationic magnetic loops of the positive defect currents and the anionic loops of the excess load negative electronic currents exert magnetic forces in opposite ways. There is always an eccentric and endocentric force, which leads to a reversal of the poles for the thermodynamic equilibrium in the interaction of geothermodynamics.

Obviously, the planet reversal magnetic field of the planets is the thermodynamic equilibrium update process. As during periods of solar flares, this weakening of the Earth's magnetic protection leaves the "open door" to the solar wind (usually deflected) and its charged particles of energy. If the northern lights and tails of comets have their share of poetry, other effects could be much more damaging to our modern societies through disturbances of electromagnetic signals. The solar system could then be

MOSTINI Planet Next Level

seen as an enlargement of the atomic level model just as the atom could be miniaturization of Solar system.

Moreover, these two sets were linked by their co-called lacunar structure (made of vacuum) even if it is more marked in the atom. These two magnitudes surround "the man, human being, living system, are integral part of the infinitely large and infinitely small as well as the Antimicrobial skin cream (microscopic and macroscopic scale.). Obviously, an atom is a miniature solar system with the nucleus in the center, the electrons in orbit and the electromagnetic force playing the role of gravity.

Here, the electrons would be kept in circular orbit with the reciprocal and perpetual of co-activities and co-interaction of the fundamental thermodynamic force and microcrack extension stress stability to the balance between centrifugal force, which tends to move them away from the nucleus, and electromagnetic force, which attracts them towards the nucleus.

MOSTINI Planet Next Level

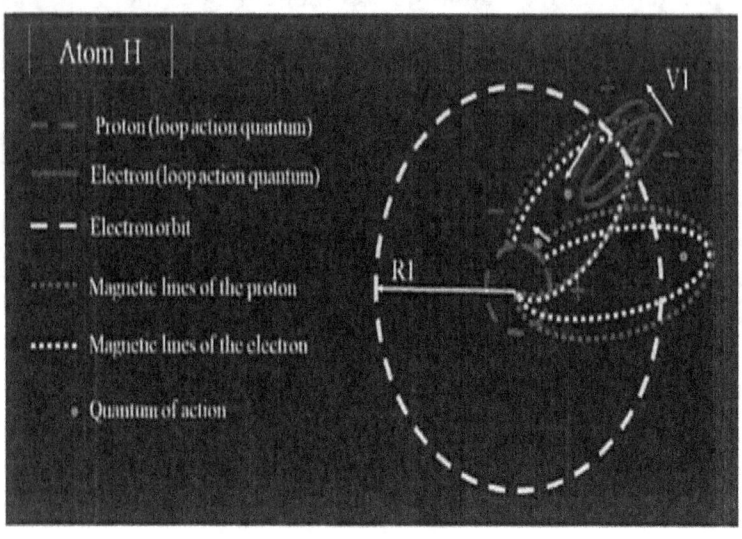

MOSTINI Planet Next Level

Chapter VI
The Fundamental Thermodynamic Force And Universe Entropy Reactivities And Functioning

Entropy Reactivities

Entropy is a disorder and defined as any system that is capable of producing energy and Geo-thermodynamic disturbance.

In the universe, the different forms of Entropy are:
- Gravitational energy
- Kinetic energy including wind energy

MOSTINI Planet Next Level

- Radiative energy including Solar energy
- Chemical energy including binding (bound), atomic, molecular, and supramolecular and fossil energy and Nuclear energy
- Electric energy

All these forms of energy are Entropy energy. Obviously, Entropic energy is proportional to the mass system, proportional light, and temperature.

Therefore, energy is stored in all the universe's entities, namely: atoms, molecules, celestials entities (planets, Stars, Galaxies, Supernovae, pulsars, dwarves, and clusters). Energy manifests itself in many ways. But whether mechanical, kinetic, thermal energy, chemical, radiation or nuclear, it can be converted from one form to another. The Entropy or Energy could use mechanical, kinetic, potential, chemical, nuclear, radial forms to do this.

MOSTINI Planet Next Level

1 Mechanical Entropy

The mechanic Entropy of object is the sum of potential and its kinetic Entropy. The mechanical Entropy is represented by Radiant Entropy that is produced by light, Electrical Entropy that is produced by electricity, Sound Entropy that is produced by sound waves, Thermal Entropy that is produced by heat.

2 Kinetic Entropy

- Kinetic entropy is the entropy of moving objects. The kinetic Entropy is proportional to the speed of this movement. The energy or Entropy of watercourse (hydraulic energy), is kinetic energy or kinetic Entropy.
- Kinetic energy can be transformed into mechanical energy or Entropy. Water mill, windmill, pump converted to a wind turbine or electricity it drives from generator.

3 Potential Entropy

The potential Entropy is the stored energy of all objects and entities in the universe. The potential energy is stored in a stationary position. We should all know that each universal object and entities has a potential energy called a gravitational centripetal energy attraction or Gravitational potential centripetal Entropy attraction. The Gravitational potential Entropy are centripetal.

This means the force of gravitational attraction is accumulating in centripetal position, in the direction of the center object Gpe (Gravitational center of object or mass: The center of gravity is the average location of the weight of an object or mass object). In the center of any universe entities, the potential gravitational force exerts an attractive and centripetal force, centripetal also in motion with centripetal force reactivities located attracting from the center of all entities. When the object or entities is in circular motion, the gravitational potential become centripetal in motion and propagates an attractive and centripetal flux, in the direction of center object. Any

MOSTINI Planet Next Level

object in the universe has in its center Gpe (the gravitational and centripetal potential entropy).

The Gpe is proportional to the mass object or entities (even a stationary object). This centripetal gravitational potential energy or entropy is present in all celestials' entities and are located in the center are centripetal in the entities as well as in the motion. When the celestial entities are in circular motion, they exert an attractive gravitational centripetal flux.

Practically, all universes entities, planets, Stars, galaxies, dwarves, clusters, pulsars, magnetars, naturals satellites, comets, presents an attraction force which originate from the center (centripetal) of entities and spread attraction from the center until surrounding objects with an centripetal in motion and in center attractive object.

When celestials entities are in motion, the flux attraction is centripetal on the Point Gpe (Gravitational potential entropy. For instance, the Sun or stars exert a gravitational attraction with all surroundings celestials entities notably all planets. The Sun's force attraction take

MOSTINI Planet Next Level

sources on the Sun centres (Gpe of Sun), and spread decreasingly from Sun through all the planets with endocentric properties and centripetal. Also, each planet exerts on the Sun and others entities an attractive and centripetal force from their centers through the Sun or surrounding entities.

The planetary attractive centripetal force takes sources from the centres (Gpe of planets) exerts on theirs surrounding notably on the Sun and other planets and attractive and centripetal in the motion as well in the entities force from their centers and during their motions exerts a centripetal force.

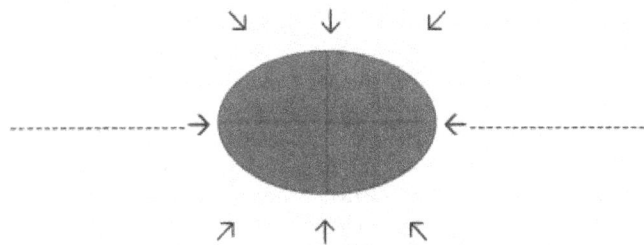

This figure gravitational centripetal attraction with

MOSTINI Planet Next Level

(Gpe) the center of gravity. Obviously, the heat energy, the radiant energy, the light energy, the electromagnetic energy, and the sound produced are Gravitational attraction centripetal force.

The Potential Thermodynamic

The potential thermodynamic is a centrifugal force,

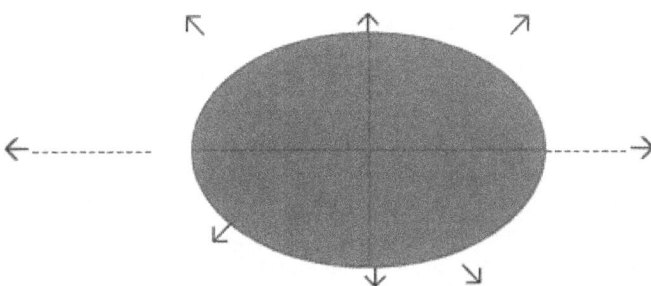

and exerts a centrifugal force to favor lowest energy all universe object, entities in the center as well as in the motion. The fundamental thermodynamic force is always centrifugal during the motion as well in the objects or celestials' entities.

MOSTINI Planet Next Level

MOSTINI Planet Next Level

The figure represents the reactivities of the fundamental thermodynamic centrifugal force. The fundamental thermodynamic centrifugal governs the universe.

MOSTINI Planet Next Level

5 Geothermal of Earth Planets (The Heat and Entropy of Earth planets)

The further one goes into the subsoil and the Earth's crust, the more the underground rocks and water became hotter and hotter, the energy of the planet increases, its Entropy increases consequently, and its gravity increases to reach an optimum point at the center Gpe (Gravitational potential entropy). The flux attraction is proportional to heat, to the mass, to the sound, and light. The attractive force is endocentric in the entities and centripetal in the planetary circular motion.

Obviously, the flux attraction is proportional originates from the center of the Earth, it is centripetal from Earth entities as well in the motion. Every planet, Solar or Stars, Galaxies, dwarves, clusters, magnetars, supernovae have their Gpe in their center and exerts centripetal motion with their surroundings entities.

In the planetary system notably on the Earth, Geothermal is a set of techniques which allow this heat to be recovered at different profounder and a different

MOSTINI Planet Next Level

temperature.

    a.   Very low temperature of geothermal energy extracts heat from the bottom at low temperature. Less than 30 degree at the depths up to 200m

    b.   The deep Geothermic of the deep Earth

The principle is the same with Geo-thermic low energy. It happens between 200 to 2500 meters. The temperature is between 30 to 90 degrees. Obviously, The further one goes into the subsoil and the Earth's crust, the more the underground rocks and water became hotter and hotter, the energy of the planet increases, its Entropy increases consequently its gravity increases to reach an optimum point at the center Gpe (Gravitational potential entropy).

The temperature on the Core is from about 4,400 degrees Celsius (7,952 degrees Fahrenheit), to about 6000 degrees Celsius (10800-degree Fahrenheit). The temperature of Earth center is the temperature Gpe (Gravitational potential entropy or energy). The Gpe is proportional to the mass entities, proportional to the

MOSTINI Planet Next Level

Entropies entities, to light entities and sound produce by entities.

We are going to calculate the forces of the gravitational pull between our plants Earth and its moon the potential Thermodynamic force.

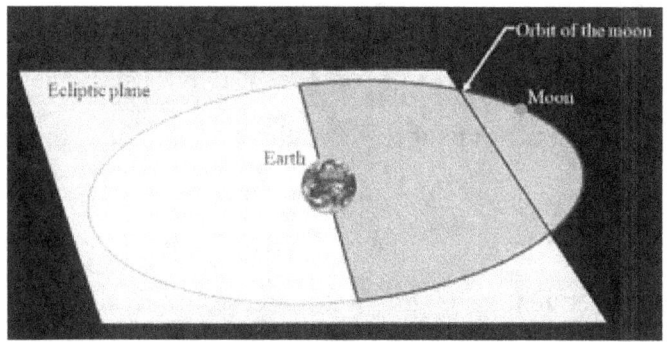

Two bodies or Entities of respective mass M1 and M2 attract each other with vectorially opposed forces of the same of the same absolute value. This value is proportional to the product of the two masses, and inversely proportional to the square of the distance between them. Each entities or body has a proper center of mass and the gravitational force is endocentric flux gravitational and centripetal during motion. From the Earth planets through the moon we

MOSTINI Planet Next Level

simulate a common and endocentric line passing the center of their proper Gravity we have:

1. For the Earth: Mass Ml, and Gpe1 (Gravity potential energy) placed in the center of corps.
2. For the Moon: M2, and Gpe2 (Gravity potential energy)

Earth

$M_1 = 5.972 \times 10^{24}$ kg

$Gpe_1 = 9,80665 m/S^2$

Moon

$M_2 = 7,3476 \times 10^{22}$ Kg

$Gpe_2 = 1,62 m/s^2$

The gravitational endocentric and centripetal force between Earth and moon is:

The line gravitational force between Moon and Earth MIM2

MOSTINI Planet Next Level

$$F_1 = F_2 = Gpet \frac{M_1 M_2}{d^2},$$

(Gpet) should be

$$Gpet = \frac{Gpe1 + Gpe2}{2}$$

$$Gpet = 6.525 \, m/s^2$$

We can make the proximity of the form of this formula on the coulomb law on the forces between —

$$|F| = \frac{1}{4\pi\epsilon_0} \frac{\Sigma Q1 \Sigma Q2}{d^{\wedge}2}$$

ΣQ = The sum of endocentric energy or heat entropy released and centripetal in motion released the Earth

$\Sigma Q2$ = The Sum of endocentric energy or heat entropy released by the moon.

The centrifugal force is a force, which moves

MOSTINI Planet Next Level

objects or celestial entities in a circular motion away from their center. On the other hand, the centripetal force is force makes a sensation in a which brings back the objects or celestials entities in circular motion movement of their center. Any force or combination of forces can cause a centripetal or radial acceleration. Just a few examples are the tension in the rope on a tether ball, the force of Earth's gravity on the Moon, friction between roller skates and a rink floor, a banked roadway's force on a car, and forces on the tube of a spinning centrifuge.

Any net force causing uniform circular motion is called a centripetal force. The direction of a centripetal force is toward the center of the curvature, the same as the direction of centripetal acceleration. According to Newton's second law of motion, net force is mass times acceleration:

Net $F = ma$. For uniform circular motion, the acceleration is the centripetal acceleration—$a = a_c$. Thus, the magnitude of centripetal force F_C is $F_C = ma_C$.

By using these expressions for centripetal acceleration ac from ac=— ac=roA2 we get two

MOSTINI Planet Next Level

expressions for the centripetal force F_C in terms of mass, velocity, angular velocity, and radius of curvature:

$$F_C = \frac{mv^2}{r};$$

$$F_C = mr\omega^2.$$

You may use whichever expression for centripetal force is more convenient. Centripetal force (F_C) is always perpendicular to the path and pointing to the center of curvature, because $\mathbf{a}_C$ is perpendicular to the velocity and pointing to the center of curvature.

Note that if you solve the first expression for r, you get:

$$r = \frac{mv^2}{F_C}.$$

This implies that for a given mass and velocity, a large centripetal force causes a small radius of curvature—that is, a tight curve.

MOSTINI Planet Next Level

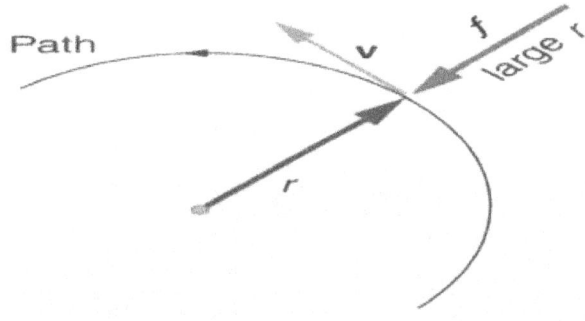

$f = F_c$ is parallel to a_c since $F_c = ma_c$

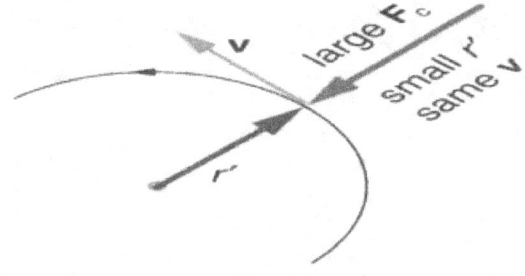

In the universe, the circular motion of any object, material or celestial entities is the set of the co-reactivities between the fundamental thermodynamic force (exocentric and centrifugal force) and microcrack extension stress stability (endocentric and centripetal), which exerts the reactivities inversely proportional in goal to maintain the object, material or Celestial entities in equilibrium position

MOSTINI Planet Next Level

zeta energy equal zero.

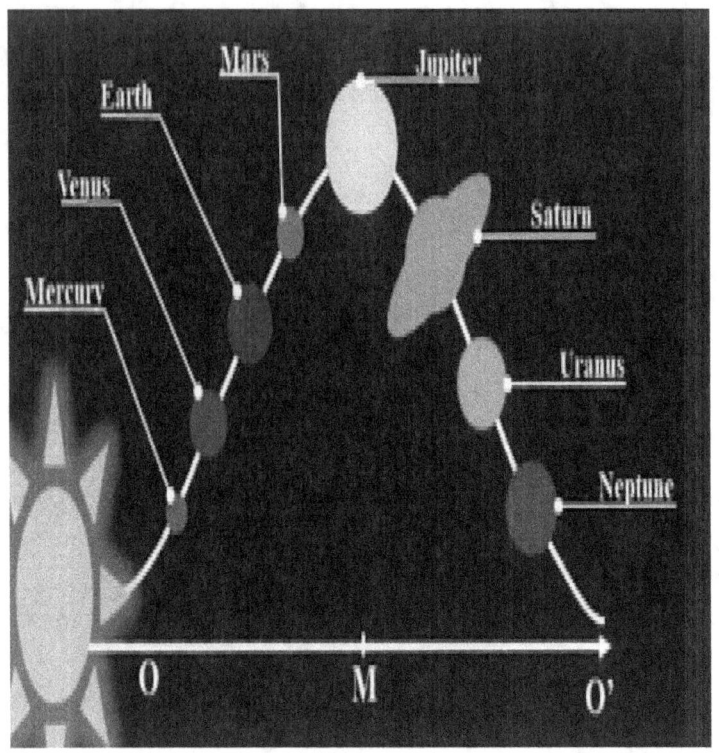

MOSTINI Planet Next Level

This figure displays the impacts reactivities and establishment by size and placement of planets in our Solar system under influence inversely proportional between the fundamental thermodynamic force and microcrack extension stress stability. Also, the gravitational centripetal force attraction level exerts between two or several entities are inversely proportional to each other, and relates their distance mass placement in equilibrium to lowest energy in equilibrium in search of zeta zero energy.

From Jupiter to Neptune

In this system, Jupiter's parameters are:

Highest gravitational force

Highest mass space entity reactivities

Highest Entropy system

Lowest thermodynamic force system

In this system Neptune's parameters are:

Lowest entropy from Jupiter to Neptune

MOSTINI Planet Next Level

Lowest gravitational centripetal force from Jupiter to Neptune

Lowest mass from Jupiter to Neptune

Highest fundamental thermodynamic force reactivity from Jupiter to Neptune

MOSTINI Planet Next Level

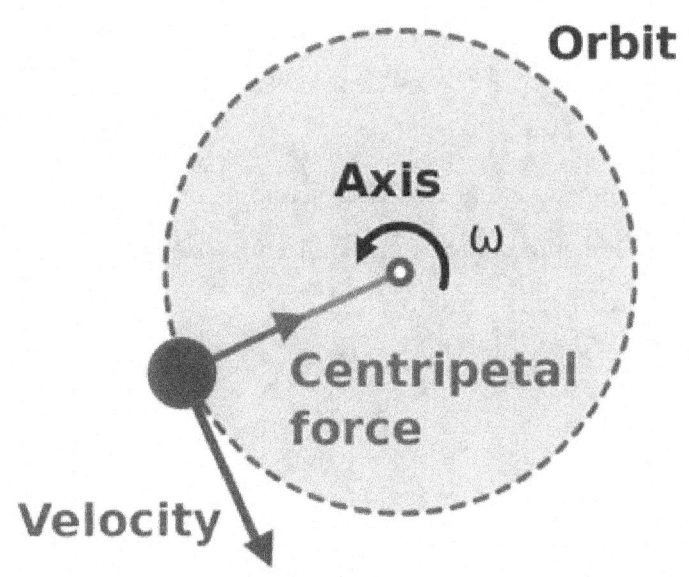

Zeta of this system in equilibrium equal zero Z=0

6 The Potential Thermodynamic Force

The Quantum Matrix Thermodynamic Sigma ZG is the establishment the interaction level of the strict

MOSTINI Planet Next Level

thermodynamic stability to the lowest energy with neatness and precision, and keen equilibrium interaction in continual, stable level correlation between the fundamental thermodynamic centrifugal force and microcrack extension stress stability centripetal force (Entropy of universe). In the universe the fundamental thermodynamic exocentric centrifugal force and microcrack extension stress stability centripetal force are inversely proportional.

The Rotation Centrifugal of Riemann

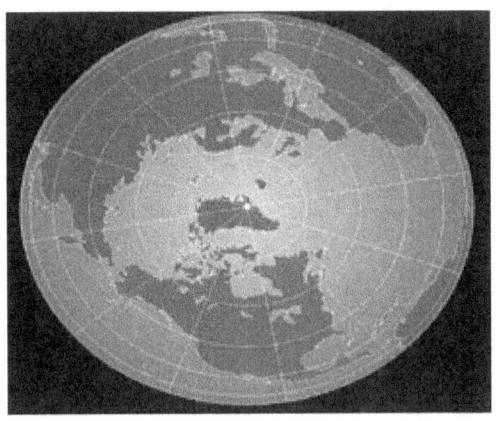

A rotation of the sphere S2 is a map r = rp, α

MOSTINI Planet Next Level

described by spinning the sphere (actually, spinning the ambient space R 3) about the line through the origin and the point p ϵ S 2, counterclockwise through angle α looking at p from outside the sphere

This reactivities of the highest cosmic law potentiate in equilibrium phase all system function of universe with respect implication reactivities of the quantum Matrix Antigravity equilibrium phase reaction with zeta=0. Planets revolve around the Sun, Galactic stars revolve around each other following the thermodynamic process, clusters, superclusters, the whole planetary system revolve, the universal hypersphere in thermodynamic equilibrium condition with zeta energy equal zero, the universe propagates. In all the universe and the universal thermodynamic hypersphere, the reactivities of the functioning are induced by two primary laws, namely:

- The fundamental thermodynamic centrifugal governs the universe and reacts inversely proportional with microcrack extension stress (the Gravitational, Entropic, and centripetal force).
- The Gravitational and Entropic force are

MOSTINI Planet Next Level

proportional to mass entities, heat, sound, and light.

These two imply the reactivities and universe functioning in equilibrium Zeta equal zero with respect of the Quantum Matrix Antigravity Sigma ZG.

Chapter VII
Thermodynamic Scale

1 Universal Thermodynamic Scale

The fundamental thermodynamic force governs the universe. The universal thermodynamic scale is based on the reactivities of the universe in Geographical areas of the universe.

2 Universe Description

The entire shape of the universe is hyper-spherical with 3D curved in 4D dimensions. Obviously, a hyper-sphere is a geometrical object in a three-dimensional space that is in the structure of a ball. Like a circle in two-dimensional space, a sphere is defined mathematically as a set of points that are all going in the same distance from a given point, but in a three-dimensional manner. The structure of the universe starts from a central point 3D. Also, our planet, solar system, Galaxies lie around the center of the entire universe (reference of thermodynamic

MOSTINI Planet Next Level

reactivities). The universal structural reactivities is not the same before and after the creation of universes entities (Malice quantum Disorder).

The universe hyper-space presents two thermodynamic phases:

- A beginning before the photoionization of the universe
- After photoionization of the malice quantum Disorder

MOSTINI Planet Next Level

3 Universe Thermodynamic Scale

The Universe thermodynamic scale begins with increasingly centrifugal universe force. The reactivities begin from the center through extremities

0 (center)—Thermodynamic increasingly centrifugal force

The Entropy of the universe decreasingly centripetal------------→

0-------------universe Entities mass special decreasingly centripetal---→

MOSTINI Planet Next Level

0----Gravitational attraction G force- decreasingly centripetal----→

3a Interpretation

As we move away from the center of the universe, the fundamental thermodynamic force increases. (lowest at 0 point and increase)

Conversely, the power of gravitational attraction decrease, which explains the expansions of universe. The Entropy of the universes from the center to the extremity. Entropy of universe is always proportional to the special Mass (planet, Galaxies, clusters, Dwarves, pulsars).

The centrifugal and increasing reactivities of the fundamental thermodynamic force (from center through the limit), justifies the hyperspherical form of universe. Obviously, other parameters to know are:

- Entropy, mass, and Gravitation with a higher value decrease from 0-point rough extremities.
- The fundamental Thermodynamic force governs

MOSTINI Planet Next Level

the universe.
- The fundamental Thermodynamic force and Entropy are inversely proportional.
- The fundamental Thermodynamic force and Gravitational energy force are inversely proportional
- The fundamental thermodynamic force is inversely proportional to the spatial mass.

3b Reactivities Thermodynamic scale of Moon, Earth planet, Jupiter, and Sun

Moon (Earth Satellite)

Sun (Our Stars)

Moon----→Earth----→Jupiter-----→Sun

a/ Decreasing of thermodynamic force reactivities (higher from the Moon and decrease through the Sun) --→

b/ The mass is increasing from the Moon through the Sun---→

MOSTINI Planet Next Level

c/ The Gravity force G increasing from the Moon to the Sun --- →

d/ Entropy force S is increasing from the Moon to the Sun ---- →

3c Interpretation

We notice that the reactivity of the fundamental Thermodynamics universe force decreases from the Moon to the Sun (higher value thermodynamic on the Moon).

Obviously, other parameters (Mass entities, Gravity, and Entropy) increase from the Moon through the Sun, so they are lower on the Moon.

4 Analyzing Thermodynamic Reactivities Of Our Solar System

If we display the size of planet from smallest to largest, we have:

Mercury->Mars->Venus->Earth->Neptune->Uranus->Saturn->Jupiter. Therefore, the biggest and largest planet is Jupiter.

MOSTINI Planet Next Level

Let's display their order in the sky to display and analyze the thermodynamic scale reactivities.

Mercury->Venus->Earth->Mars->Jupiter->Saturn->Uranus->Neptune.

4a Analyzing Thermodynamic Distance Between Our Sun And Jupiter

The Geo-thermodynamic is the ultimate tool to analyze universe reactivities. From the Sun through Mercury, as the Entropy decreases, the mass decreases, and the gravitational force decreases. On the other hand, the fundamental thermodynamic force increases this justification of establishment and the presence of Mercury. Increasingly the presence of the fundamental thermodynamic force justifies at 58 million kms the presence of Mercury and Venus at 108 million kms in conclusion, the establishment of reactivity between the two or several celestial entities are controlled by the fundamental thermodynamic force.

Sun--------------------------------→Mercury

MOSTINI Planet Next Level

From Sun to Mercury, the gravitational centripetal attraction force is high and impact on mercury size. Mercury size is small because the influence of highest gravity exerts by the Sun. Also, Mercury is coated with level of the fundamental thermodynamic force. The fundamental thermodynamic centrifugal is inversely proportional to mass entities.

- The mass entities reactivities decreases
- The Entropy reactivities decreases
- The gravity force decrease
- The fundamental force thermodynamic Increase.

They represent the same thermodynamic parameters

- From Sun to Jupiter - From Sun to Neptune

4b Analyzing Thermodynamic Reactivities between

- Mercury and Jupiter
- Mercury->Venus->Earth->Mars->Jupiter

MOSTINI Planet Next Level

- The Mass of System increase from mercury to Jupiter----→
- The Entropy of System increase from mercury to Jupiter----→
- The Gravity force system increase from mercury to Jupiter ----→
- The Thermodynamic decrease from mercury to Jupiter---→

In this system, Mercury's parameters are:

- Lowest level of gravitation force
- Lowest level of mass space entity
- Lowest level Entropy

Highest level of the fundamental thermodynamic force. Jupiter's parameters are:

- Lowest fundamental thermodynamic force
- Highest gravitational force G
- Highest Mess space entity
- Highest Entropy

MOSTINI Planet Next Level

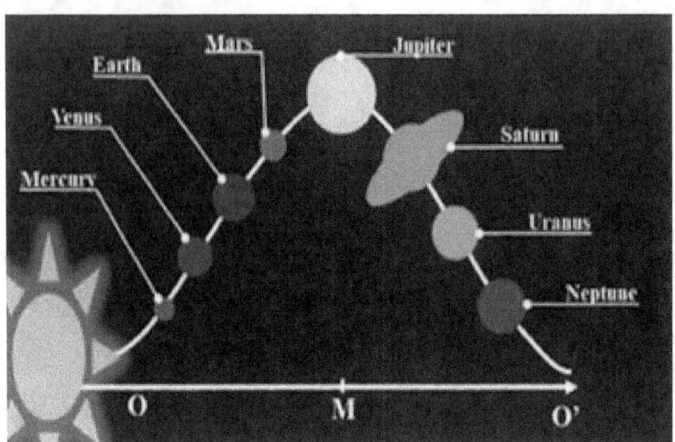

4c Analyzing reactivities between Jupiter and Neptune

Jupiter→Saturn→Uranus→Neptune

The mass space entity decreases from Jupiter to Neptune.

The Entropy decrease from Jupiter to Neptune.

The Gravity attraction decrease from Jupiter to Neptune.

Increase of the fundamental thermodynamic force from Jupiter to Neptune.

In this system, Jupiter's parameters are:
- Highest gravitational force

MOSTINI Planet Next Level

- Highest mass space entity reactivities
- Highest entropy system
- Lowest thermodynamic system

In this system, Neptune's parameters are:

- Lowest mass from Jupiter to Neptune
- Lowest Entropy from Jupiter to Neptune
- Lowest gravitational force G from Jupiter to Neptune
- Highest thermodynamic force from Jupiter to Neptune

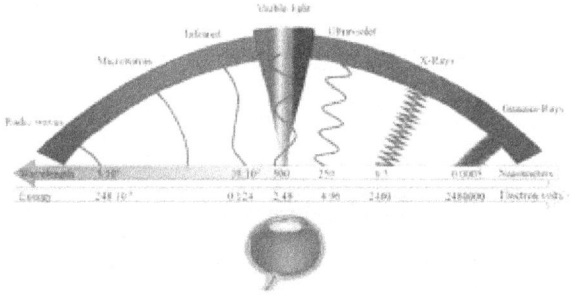

This figure displays the reactivities of planets alignments with mass, entropy, gravity force, and fundamental thermodynamic force proportion implication from Sun (Stars) to Mercury, from Mercury to Jupiter, from Jupiter to nature. Obviously, from Mercury to Neptune, if we inverted this path reactivities with X planets in continual inversion conforming reactivities of figure above,

MOSTINI Planet Next Level

we are going to have a second, third, fourth solar system with planets X.

We have:

Solar system (our systems) + Solar System (X planets) + Solar System (X' planets) + Solar System (X" planets)

We are going to have, Galaxies, clusters, superclusters.

5 Quantum Entities Identification (QEI)

QEI consists and help to identify the safety of planet from Geothermodynamic, Entropy, and gravitational reactivities. Also, the QEI method helps to identify the distance of exoplanets, Exostar, identical reactivities exoplanets, twin planets or sister planets (two planets identical with the same reactivities but separate in the geographical environment, two identical stars separate each one by separate distance with the same reactivities (twin stars).

MOSTINI Planet Next Level

5a Security Identification

The safety of a planet is determined by the gravity constant, the more the thermodynamic constant is high, the more the gravity is weak. Let's compare safety between the moon our satellite, our planets, and Jupiter.

1) Gravity   Thermodynamic
   Moon= 1.625 m/s^2 0.615s^2/m
   Earth= 9.80 m/s^2 O. 102s^2
   Jupiter= 24.9 m/s^2 0.00401 s^2/m

The moon has the lowest gravity and the highest thermodynamic force, so the moon is the safest entity between all of them. The second is Earth with 9.8 m/s^2 and 0.102s^2/m. The least safe is Jupiter, with the highest gravity and the lowest thermodynamic. Jupiter is not safe. Our moon is very well protected, but it has traces of impacts from craters, so in view of its thermodynamic conditions. These impacts are caused by mass objects larger that may be half or double of its size.

MOSTINI Planet Next Level

5b Identification of Entities by the QEI method

The determination and identification of exoplanets, Exostar, twin plants, and sister planets is done by symmetry inverse of topology placements of wavelengths structure and path topology. This topology placement is the thermodynamic holarchy and centrifugal reactivities. In the universe the size of celestial entities (planets, Stars, dwarves, pulsars, magnetar) are always determined by the co-reactivities between the fundamental thermodynamic centrifugal inversely proportional with the microcrack extension centripetal force stress stability in the hyperspace holarchy cosmic universal. The Stars, celestial entities never develop at random. There is a holarchy (Holon) reactivity harmony governed by the universe structure, the fundamental thermodynamic centrifugal force in co-reactivities is inversely proportional with microcrack extension stress stability centripetal force. The celestial disposition entities are inversely symmetrical in the universe's spherical holarchy area.

MOSTINI Planet Next Level

The celestial's entities are always identical to the topology path of wavelength in planetary size, following the centrifugal direction of the fundamental thermodynamic force reactivities. The celestial entities never place themselves at random, and never develop at random, there is holarchy harmony (Holon) of the arrangements, formation, and development of planets and stars according to the fundamental thermodynamic centripetal force reactivities and the topology of waves. Obviously, in the universe, everything is organized and reacts in accordance with the Geo-thermodynamic reactivities with accuracy and with strict precision. The structure organizational of universe is a Holon structure. We have abilities with method QEI method to detect, to establish size, and identify planets, galaxies, clusters, size precise, and distance with precision and the Geothermodynamic structure of any part of our universe.

Mars should be bigger than Earth but is very small because of higher gravity of Jupiter. Obviously, Jupiter exerts high gravity on Mars. Mars is small with high level thermodynamic law. The fundamental thermodynamic

MOSTINI Planet Next Level

centrifugal force governs the world. The fundamental thermodynamic centrifugal and attraction gravitational centripetal force are inversely proportional

6 The Folding Universe

This technic consists of bending the universe when we imprint the Solar axes in the Galaxies in a repetitive way under these conditions where the distances of planetary alignments in geometric form are like wavelengths features.

With this technic we reach new planets, Solar system, galaxies, and clusters within several light years, this technic is called "Josammy Emporio 3D" transmigration technology.

7 Mars Planets Irrelevance Thermodynamic Status

Mars planet lost its internal energy billions of years ago, therefore it does not have a magnetic fields nor

MOSTINI Planet Next Level

internal electricity. Mars lost global magnetic and water billions of years ago. Mars is a dead planet because thermodynamically its position is irrelevant. Mars functions with low energy, which is why it occupies a Thermodynamically irrelevant position.

Currently its orbit is ensured by a very important value of the fundamental thermodynamic force. Mars is an ambiguous planet with very low internal energy. All planets in the universe renew and reactivate their kinetic and potential energies by magnetic inversion, but in the case of Mars, it is improbable. The planet Mars would be disoriented later by gravity of Jupiter.

The thermodynamic order of the planets should be distributed as follows:

Mercury->Mars->Venus->Earth->Jupiter->Saturn->Uranus->Neptune

7a Interpretation

Potential energy is stored in all objects, atoms,

MOSTINI Planet Next Level

molecules, planets, Galaxies, Stars, Solar systems, Supernovae, pulsars, clusters. Obviously, we notice with reactivities of Mercury, Mars, Venus, Earth, and Jupiter that the increasing size, that should be the increasing of Entropy and potential energy. Also, the decreasing size from Jupiter through Neptune, we notice the decreasing size with decreasing reactivities of entropy and potential energy.

This is thermodynamic universal planet placement reactivities. However, in our Solar system, the planet, Mars presents an ambiguity because of its irrelevant place. Mars lost its internal, the centripetal entropy which is gravity about 5 billion years ago. The core of Mars begun to cool, and progressively lost its internal centripetal heat, which is its gravitational force. Therefore, the gravitational force of Jupiter followed by the rank of Saturn and Uranus attracted Mars to retrieve its actual orbital. If Mars continues to lose its internal centripetal energy, we will be in the future, the natural satellite of Jupiter without any magnetism and internal energy. Mars is a dead planet; a cemetery. Thermodynamically it is irrelevant with an inappropriate position. The consequence of ambiguity is that the very slower clockwise rotation (speed rotation of Venus) of Venus in our solar system. Venus rotates clock-wise in retrograde once every 243 Earth days; the lowest rotation of any planet. On the contrary, all planets rotate on their axis in an anti-clockwise direction. Potential energy is stored in all objects, atoms, molecules, planets, Galaxies, Stars, Solar systems, Supernovae, pulsars, clusters.

Obviously, we notice with the reactivities of Mercury, Mars, Venus, Earth, Jupiter, the increasing size, that should be the increasing of Entropy and potential energy.

8 The Mathematical Language Of Universal Reactivities

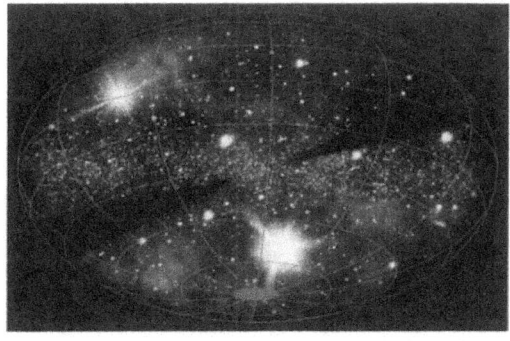

MOSTINI Planet Next Level

The mathematical implication reactivity of the fundamental thermodynamic force refers and elucidates the

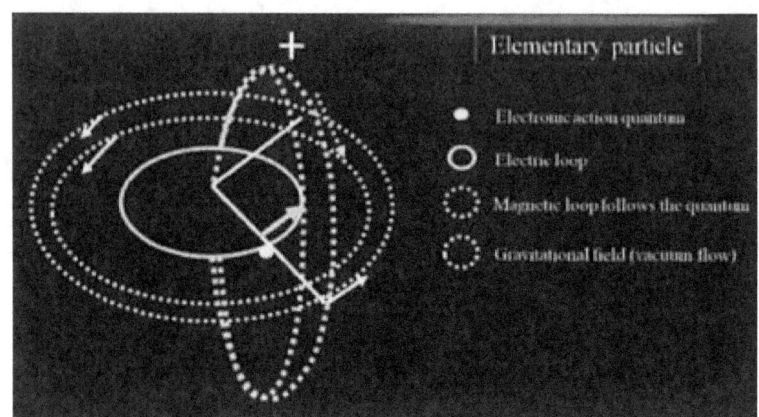

parameters characteristic of universe functioning, these characteristic parameters are: reactive(Sigma ZG), oxidoreduction and Geo-thermodynamic or geographic regulation), structural, quantitative, and evaluation Symbolic.

8a Structural

Potential energy is stored in all objects, atoms, molecules, planets, Galaxies, Stars, Solar systems, Supernovae, pulsars, and clusters.

Obviously, we notice with the reactivities of Mercury, Mars, Venus, Earth, Jupiter, Saturn, and Uranus

MOSTINI Planet Next Level

that the increasing size, that should be the increasing of Entropy and potential energy.

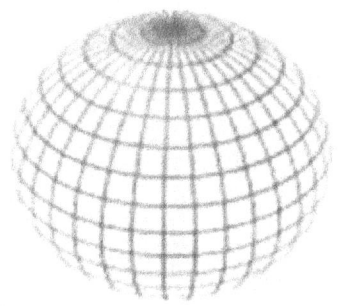

The Zeta = O

All surface is surrounded by negative charge pressure of Dark energy. Dark energy is a universal force that exerts negative pressure in all entities. The negative pressure begins slightly from point C (center) and leading toward surface on a growing scale of Thermodynamic reactivities.

MOSTINI Planet Next Level

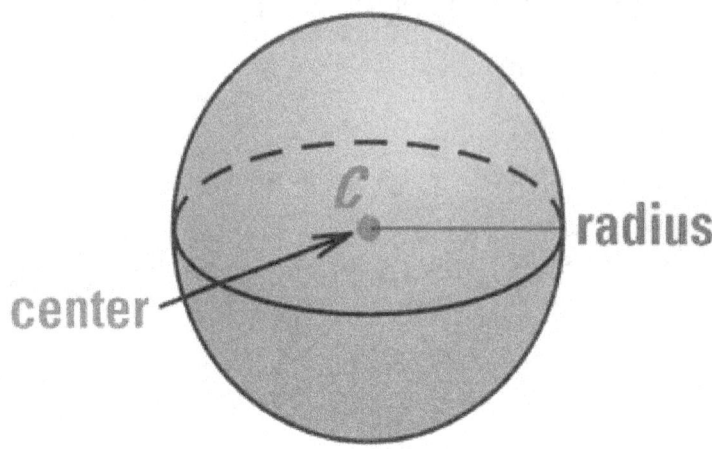

All surface is surrounded by negative charge pressure of Dark energy. Dark energy is a universal force that exerts negative pressure in all entities. The negative pressure begins slightly from point C (center) and leading toward surface on a growing scale of Thermodynamic reactivities

MOSTINI Planet Next Level

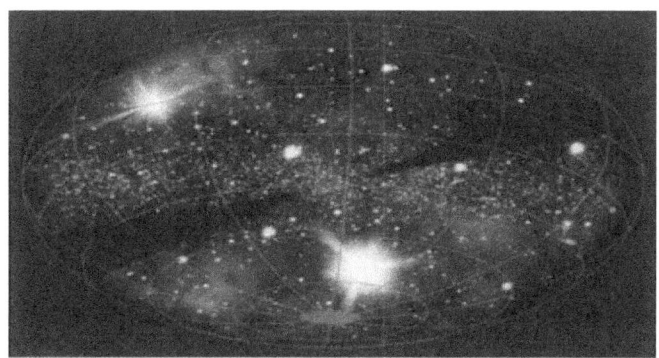

All surface is surrounded by negative charge pressure of Dark energy. Dark energy is a universal force that exerts negative pressure in all entities. The negative pressure begins slightly from point C (center) and leading toward surface on a growing scale of Thermodynamic reactivities

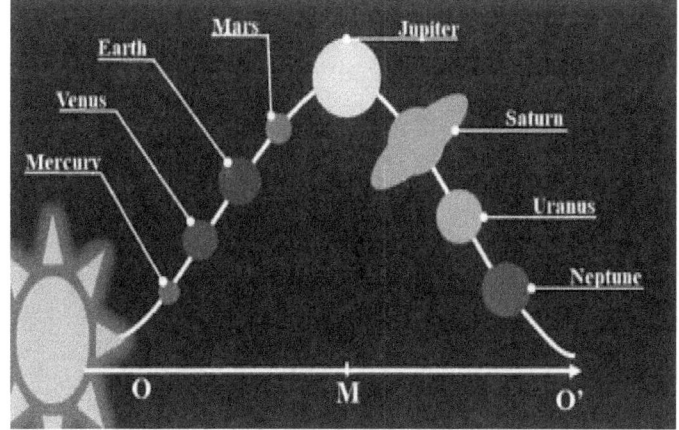

MOSTINI Planet Next Level

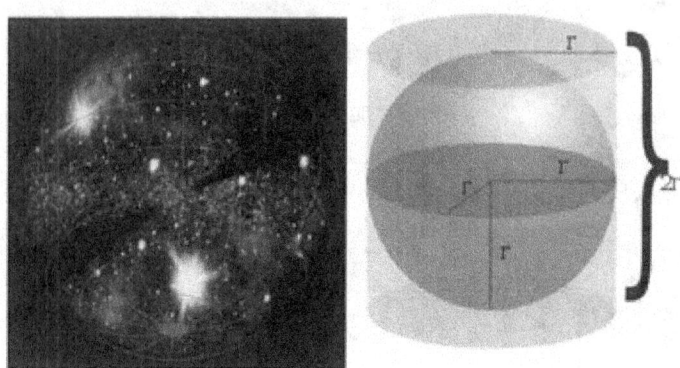

The universe is a spherical hyperspace whose structural form is represented in 3D dimensions. The universe is not infinite, but it is finite with a spherical circumference in 3D curved in 4D. All the surface is surrounded in negative charge pressure of Dark energy. Dark energy is a universal force that exert negative pressure in all entities. The negative pressure begins slightly from point C (center) and leading toward surface on a growing scale of Thermodynamic reactivities.

The entire shape of universe is spherical with 3D dimension curved in 4D. A sphere is a geometrical object in a three-dimensional space that is structure of a ball. Like a circle in two-dimensional space, a sphere is defined

MOSTINI Planet Next Level

mathematically as a set points that are all the same distance from a given point, but in a three dimensional one. The structure of the universe starts from a central point 3D. Also, our planet, solar system, Galaxies and more lie around the center of the entire universe (reference of thermodynamic reactivities). Universal structural reactivities is not the same before and after the creation of the universes entities (Malice quantum Disorder).

9 The Universe Primordial Thermodynamic Reactivities

1) In the primordial universe before the creation or the beginning of Malice Quantum Disorder. The universe did not have any Entropy (Zeta—O).

$$\lim_{T \to 0} (\Delta S)_T = 0$$

The physicochemical potential was zero, T = 0. These physical characteristics represented the highest and strongest of the fundamental Thermodynamic force, which

MOSTINI Planet Next Level

is dark energy exercising negative pressure in all universe. Obviously, the universe was like a great primordial era and insulating composed and fulfilled with Dark Energy (the fundamental primordial universe) interrelated with isolated elements to know: Matter and Anti-matter. In the beginning, the structure of the universe was the Dark Energy era, or the fundamental Thermodynamic primordial force that fulfilled all the universe, interrelated, and dotted with practically equal amounts of matter and antimatter. The universe of the whole system was zero, temperature zero, Entropy zero. In the primordial universe, there wasn't magnetism, radioactivity. However, there was anti-matter, matter, and proton in the thermodynamic reactivity equilibrium of the high level of the fundamental thermodynamic force or Dark energy with black in color. The matter and antimatter were in equilibrium and purity with straight-line configuration.

2) The structure of the universe after entities creation (Malice quantum disorder and photoionization) changed because of thermodynamic instability, the photoionization and malice quantum disorder took effect,

MOSTINI Planet Next Level

when dark energy or the fundamental thermodynamic regulate energy to conform in the condition of universe primordial , another problem arise relative to the first problem solved. The fundamental thermodynamic and Entropy are inversely proportional. Moreover, the statistic which reacts to the spacings between zero of Riemann's function is exactly the same as the statistic govern the energy levels of the theoretical random atoms. If we build an atom with random bricks, we measure energy levels, and we will find the statistics of Riemann.

We have connection and correlation between quantum mechanics and prime numbers. The difference and between the zero of Riemann is the as the statistics of the energy levels which exist at the level of the random atoms. The fundamental Thermodynamic force governs the world. We have a representation curvature, which represent Riemann, Gauss, Boltzmann, Einstein, RICCI, Kepler, Nicolas Tesla, Avogadro to describe the relationship between Entropy and disorder in a system. We have $S=K\ln W$

The Riemann zeta function is defined for complex s

MOSTINI Planet Next Level

with real part greater than 1 by the absolutely convergent infinite series Leonhard Euler He also proved that it equals the Euler product where the infinite product extends over all prime numbers p. The Riemann hypothesis discusses zeros outside the region of convergence of this series and Euler product. To make sense of the hypothesis, it is necessary to analytically continue the function to obtain a form that is valid for all complex s. This is permissible because the zeta function is meromorphic, so its analytic continuation is guaranteed to be unique and functional forms equivalent over their domains. One begins by showing that the zeta function and the Dirichlet eta function satisfy the relation. In this universe, we have a new configuration of universe deformation.

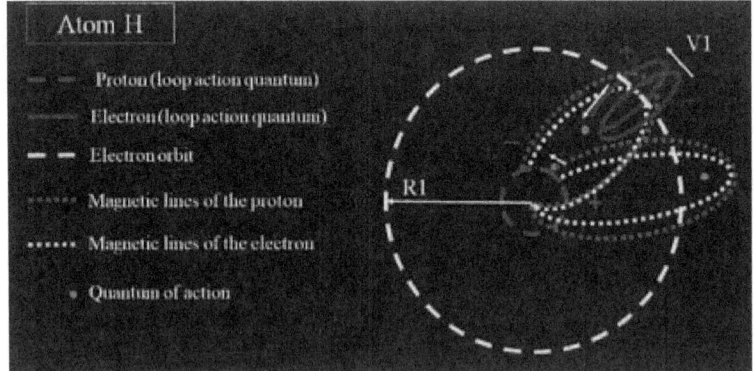

This configuration displays the reactivity of the

MOSTINI Planet Next Level

universe after photoionization, which is the Quantum Malice Disorder, configuration displayed in the configuration reactivities of infinitely small and infinitely large.

10 The Geo-thermodynamic Form Reactivities

The fundamental thermodynamic force governs the world. The fundamental thermodynamic force and microcrack extension stress stability (Entropy) are inversely proportional. The Entropy is proportional to the mass system. The Gravity is inversely proportional to the fundamental thermodynamic force. Therefore, Geo-thermodynamic reacts by flux thermodynamic is reacting horizontally, vertically, and obliquely. The Geo-thermodynamic reacts always to favor energy in the geographical area. The fundamental Thermodynamic centrifugal force may act in all directions namely: horizontal, vertical, and obliquely to favor energy (thermodynamic flux) reacts inversely proportional in all

MOSTINI Planet Next Level

direction with Centripetal Entropy to favor all energies (Entropy flux). The co-reactivity is inversely proportional between the fundamental thermodynamic and microcrack extension stress always favors reactivities to zeta =0, and governs and ensures the reactivities of planets rotation, Galaxies, and all universe.

The fundamental Thermodynamic force (Geo-thermodynamic) with its reactivities inversely proportional govern the planets ensuring rotation around them and around their orbit.

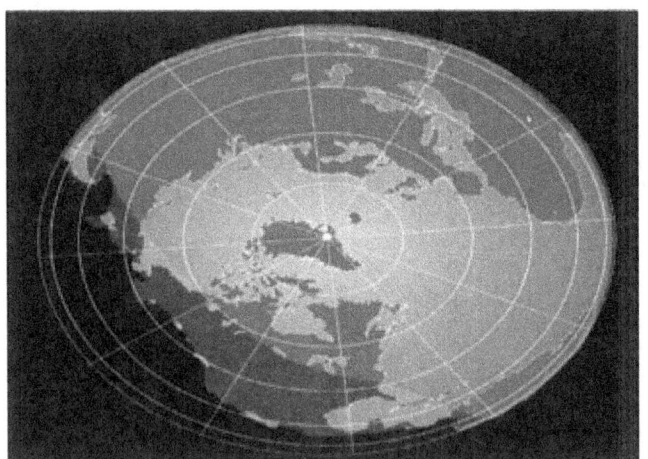

Potential energy is stored in all objects, atoms, molecules, planets, Galaxies, Stars, Solar systems,

MOSTINI Planet Next Level

Supernovae, pulsars, and clusters.

Obviously, we notice with the reactivities of Mercury, Mars, Venus, Earth, Jupiter, Saturn, Uranus, the increasing size that should be the increasing of Entropy and potential energy.

11 The Universal Geo-Thermodynamic Scale Reactivities

a) The scale of the universal thermodynamic reactivities is the evaluation and the potential reactivities path.

The equilibrium positions of atoms and molecules in the condensed phase are the result of an interplay between the forces of repulsion and attraction, and the motion in substance (the reactivities of the fundamental force and Entropy). This explains why thermodynamic consideration can make important contribution to theories of the structure matters. Advance in the knowledge of various crystal structure have, on the other hand, led to considerable progress in thermodynamic. We could compare this Thermodynamic reactivities process with electronegativity process in the periodic table.

Electronegativity is the tendency of an atom or molecule to attract electrons towards itself. Electronegativity generally increases across periods. The atoms which are electron negatives are great oxidant (Oxygen, fluoride). Electronegativity increases from left to right across the periodic table in the order C<N<O<F. In the universe. In the universe, the fundamental thermodynamic force increases from the center through the limit surface of universe sphere in all directions of universe limit sphere.

12 Oxidoreduction Reactivity Through Universe Reactivities

The highest electronegativity the greater pull an oxidizing agent. The fundamental thermodynamic force govern universe, the fundamental and microcrack extension stress stability are inversely proportional (Entropy). The Entropy is always proportional to the mass entities. The

MOSTINI Planet Next Level

fundamental thermodynamic force is inversely proportional to the mass entities of system in the universe.

The evaluation sigma ZG charge of universe entities are always related to thermodynamic mathematic evaluation know:

By considering the scale of oxidizing agent in a periodic table to know:

K, Ca, Na, Mg, Al, Mn, Zn, Fe, Pb, H, Cu, Ag, Hg

The electro positivity increases as we move from right to left in the periodic table. Hence, the oxidation number/oxidizing power increase as we move from left to right in the modern periodic table. All these elements will be positive when they are going the lose electrons. Also, if we consider reactivities between Ag^+ / Cu and between Cu^{2+} /Fe we have

MOSTINI Planet Next Level

$Ag^+ + e \longrightarrow Ag(s)$ (Reduction)
$Cu(s) \longrightarrow Cu^{2+} + 2e$ (oxidation)
$2Ag^+ + Cu(s) \longrightarrow 2Ag(s) + Cu^{2+}$

$Cu^{2+} + 2e \longrightarrow Cu$ (reduction)
$Fe \longrightarrow Fe^{2+} + 2e$ (oxidation)
$Cu^{2+} + Fe \longrightarrow Fe^{2+} Cu$

The couple Cu2+/Cu present an ambivalent character in the measure of reacting as an oxidant or reducer, Ag is more oxidizing than cooper. Also, cooper is more oxidizing than iron. Therefore, Cooper can be oxidizing or reducing depending on the periodic table oxidation scale of element. In both reactions, the copper has two character, we have oxidation and reduction.

Cu2+/Cu

We have:

$Cu^{2+} + 2e \longrightarrow Cu$ (Reduction)
$Cu \longrightarrow Cu^{2+} + 2e$ (oxidation)
$Cu^{2+} + Cu \longrightarrow Cu + Cu^{2+}$

In the thermodynamic concept reactivities, in infinitely large as well as infinitely small, the reduction

MOSTINI Planet Next Level

implies negative charge (-) while oxidation implies (+) charge. This reactivities characters are similar to the reactivities of planets, Galaxies, solar systems, clusters with the concept sigma reactivities Σ in the universe and all celestial entities. All celestials' entities endowed with a high thermodynamic value or high charge thermodynamic have a great and high attractive influence. Therefore, they are less attractive Σ. On the other hand, all celestial entities endowed with low value by comparison thermodynamic, with low value or low charge thermodynamic are more attractive Σ (+)

13 Comparison of the Moon (Earth Satellite), the Planet Earth, Jupiter, Saturn and Stars (our Sun)

The Earth planet is bigger than the moon, so the Entropy of Earth is higher than the moon.

Because the Entropy is bigger on Earth, the Entropy is always proportional to the mass, so the mass of the Earth is bigger and the fundamental thermodynamic is lower than

MOSTINI Planet Next Level

in the moon. Attraction sigma charge on this is lesser $\Sigma(-)$ than in the Earth Σ (+). The value of thermodynamic is less than on moon. The Earth by comparison attraction sigma Σ between Earth and Jupiter.

Attraction sigma charge of Earth is lesser Σ (-) by comparison attraction of Jupiter high Σ (+) because in Jupiter the Entropy is higher than Earth, the mass is huge the thermodynamic force on Jupiter is lesser than Earth.

1) Moon attraction charge is lesser Σ (-) than Earth. Earth charge attraction is higher Σ (+), Entropy system value is higher and mass by comparison with moon. The fundamental thermodynamic force reactivity on Earth is lesser in Earth than moon.

2) Earth's attraction charge is lesser Σ (-) than Jupiter. Jupiter attraction is higher $\Sigma(+)$, Entropy system value is higher and mass by comparison with Earth. The fundamental Thermodynamic force reactivities on Jupiter is less than on Earth

3) Jupiter's attraction charge is lesser $\Sigma(-)$ than Sun star. Sun stars attraction is higher $\Sigma(+)$, Entropy is higher

MOSTINI Planet Next Level

and mass by comparison with Jupiter. The fundamental thermodynamic force in the Stars is less than Jupiter.

Any celestial entities could have $\Sigma(+)$ or $\Sigma(-)$

By conclusion, the reactivities of the fundamental thermodynamic force determines the sigma value reactivities of celestial's entities. The higher reactivities of the fundamental force, the lower the attraction charge, the system Entropy, the low mass value. On the other hand, lower reactivities of the fundamental thermodynamic force, the higher Entropy system, higher mass value. The fundamental thermodynamic force governs universe. The Entropy is proportional to the mass and inversely proportional to the fundamental thermodynamic force.

14 Geo-thermodynamic rotation charge identification

In the geographical environment, the fundamental thermodynamic force always favor reactivities.

The universe is a spherical hyperspace whose structural form is represented in 3D dimension curved in

MOSTINI Planet Next Level

4D. The universe is not infinite, but it is finite with a spherical circumference in 3D curved in 4D. All surface is surrounded in negative charge pressure of Dark energy. Dark energy is a universal force that exert negative pressure in all entities.

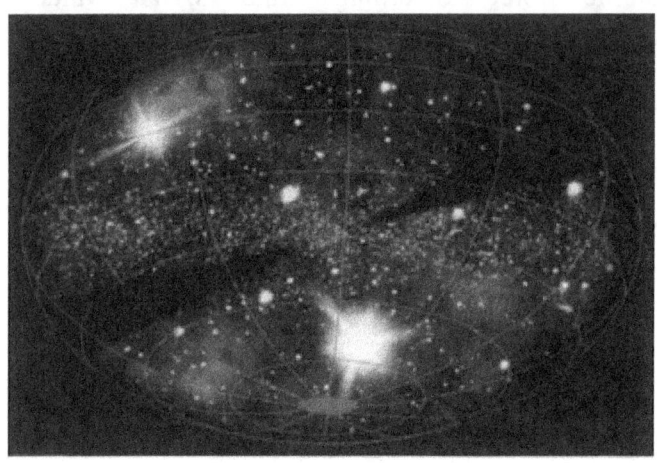

The figure above is representing the universe in a microscopic scale. This representation is similar to the representation of infinitely large.

We have structural similarities between atoms, planets, stars, and the entire universe. Obviously, the potential energy is stored in all objects, atoms, molecules,

MOSTINI Planet Next Level

planets, Galaxies, Stars, Solar systems, Supernovae, pulsars, clusters. Obviously, we notice with the reactivities of Mercury, Mars, Venus, Earth, Jupiter, Saturn, Uranus, the increasing size that should be the increasing of Entropy and potential energy.

In the universe, the Geo-thermodynamic reactivities establish always sigma charge reactivities and establishes interaction and reactivities in the proper thermodynamic mathematic as follow: zero is the universe center $\Sigma(-1)$, $\Sigma(-2)$, $\Sigma(-3)$, $\Sigma(-4)$, $\Sigma(-5)$, $\Sigma(-6)$, $\Sigma(-7)$,

14 a. Quantitative

By comparison we have:

$(-1, 0)$, we have $\Sigma(-1) = [-], \Sigma(0) = [+]$
$(-1, -2)$ we have $\Sigma(-2) = [-]$. $\Sigma(-1) = [+]$
$(-2, -3)$ we have $\Sigma(-3) = [-], \Sigma(-2) = [+]$
$(-4, -3)$ we have $\Sigma(-4) = [-], \Sigma(-3) = [+]$
$(-5, -4)$ we have $\Sigma(-5) = [-], \Sigma(-4) = [+]$
$(-6, -5)$ we have $\Sigma(-6) = [-], \Sigma(-5) = [+]$
$(-7, -6)$ we have $\Sigma(-7) = [-], \Sigma(-6) = [+]$
$(-8, -7)$ we have $\Sigma(-8) = [-], \Sigma(-6) = [+]$
$(-9, -8)$ we have $\Sigma(-9) = [-], \Sigma(-8) = [+]$

MOSTINI Planet Next Level

This reactivity is operating in all hyper spherical universe, favoring all universe reactivities functioning including the universe expansion.

The formation of planet, Solar system, Galaxies, dwarves, Supernovae.

$(-10, -9)$ we have $\Sigma(-10)=[-], \Sigma(-8)=[+]$

In the geo-thermodynamic reactivities, Celestial entities may have same parameter, the positive (+) charge is assigned to the weakest charge attraction. For instance: with thermodynamic reactivities.

The fundamental thermodynamic force govern universe, the fundamental and microcrack extension stress stability are inversely proportional (Entropy). The Entropy is always proportional to the mass entities. The fundamental thermodynamic force is inversely proportional to the mass entities of system in the universe. The evaluation sigma ZG charge of the universe entities are always related to thermodynamic mathematic evaluation know: Sigma ZG reactivities of universe

MOSTINI Planet Next Level

In the hyper-spherical finite universe and huge universe, the reactivity of planet, Stars Solar, system, Galaxies, dwarves, supernovae are induces and implements by the fundamental sigma ZG which control all universe characterizing by sigma ZG of the fundamental thermodynamic force giving Σ(entitie) =+ or Σ (entity)= -

In the universe all stars are aligned with centrifugal thermodynamic increasing level position with an alternation of sigma charges which constitutes a great thermodynamic equilibrium and all structure form a Galactic spiral.

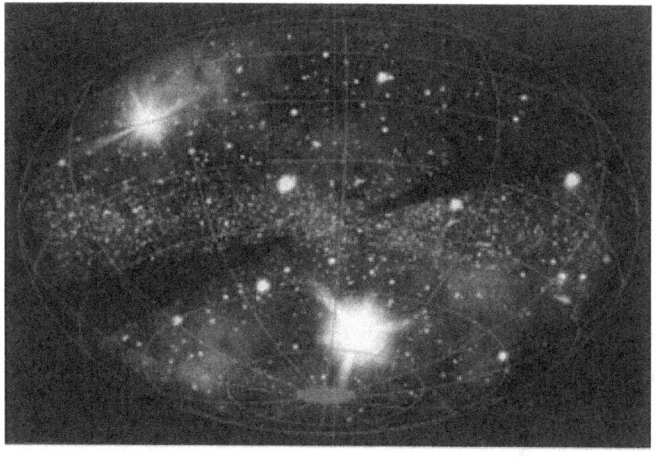

Potential energy is stored in all objects, atoms,

MOSTINI Planet Next Level

molecules, planets, Galaxies, Stars, Solar systems, Supernovae, pulsars, clusters. Obviously, we notice with reactivities of Mercury, Mars, Venus, Earth, Jupiter, Saturn, Uranus, the increasing size that should be the increasing of Entropy and potential energy.

15 Quantum universal Entities Reactivities (QUER)

15a The rotation of Celestial entities around Themselves

In the universe, all celestial entities react and interfere by respecting and conforming the reactivities of highest cosmic laws namely:

- The fundamental thermodynamic force governs universe.
- The fundamental thermodynamic and microcrack extension stress stability (Entropy) are inversely proportional.
- The Entropy of system is proportional to mass system

MOSTINI Planet Next Level

- The Entropy is proportional to gravitational force

The gravitational force, Entropy, and mass entities are inversely proportional to the fundamental thermodynamic force. Therefore, the rotation of celestial entities (planets, Stars, dwarves) around themselves is the co-reactivities inversely proportional between the fundamental thermodynamic and gravitational giving continual reactivities rotations in equilibrium condition of Zeta = zero. We may display reactivities of the rotation of planets, Stars, dwarves. Solely planets or Stars or dwarves the principle of the Riemann 'Rotation Sphere. Also, the fundamental thermodynamic always exerts a centrifugal force in all universe entities to favor reactivities and reactions. This centrifugal force can be seen in the formula - $F=m\omega^2 r$

This centrifugal thermodynamic force is at the origin of the flattening of celestial entities on the axis of the poles and conforming all universe entities on the position of lowest energy and the round shape of celestial entities.

The fundamental thermodynamic exercising a

MOSTINI Planet Next Level

centrifugal force while the gravity is centripetal and resulting of round shape and reactivities to lowest energies in the goal to favor interaction with celestial entities. The fundamental thermodynamic force is always centrifugal while gravity is centripetal and causing rotational movement around themselves by exercising a negative pressure in all inverse entities in infinitely small and infinitely large, in the microscopic scale as well as macroscopic scale and the planetary zeta =0. Obviously, the reactivity electron around maintained around fundamental orbital is the reactivity of the fundamental thermodynamic force with zeta atomic =0. For the Sun, the huge celestial entities have on force centrifugal (the fundamental thermodynamic force) in equilibrium with gravitational centripetal force generating round shape Solar Corona with Zeta = 0.

16 The universal force reactivities scale

The universal composition scale reactivities are:

- The fundamental thermodynamic centrifugal

MOSTINI Planet Next Level

   force
- The Gravitational G centripetal force reactivities proportional to mass and Entropy
- The Entropy force reactivity.

Their scale should be:

- --→
- The centrifugal thermodynamic force increasing reactivities
- --→
- The centripetal Gravitational force decreasing reactivities
- --→
- The centripetal Entropic force decreasing reactivity

17 Interpretation

According geo-thermodynamic reactivities sigma ZG, Mercury is planet is the smallest planet in our solar system and has the highest thermodynamic force reactivities. Although, Mercury is very close to the Sun, the

MOSTINI Planet Next Level

centripetal force gravity never affects it to rotate in the in opposite direction clockwise like all planets in our Solar. However, in Mercury the centrifugal fundamental thermodynamic force is higher and challenge the centripetal and gravitational force of Sun. Most planets will turn counterclockwise as well as mercury, with the exception of Uranus, Neptune and Venus.

Obviously, Venus is unusual because it spins in the opposite direction of Earth and most other plants. And it rotates very slow. It takes about 243 Earth days to spin around just once. Because it's so close to the Sun (Centripetal Gravitational force). A year goes by fast. it takes 225 Earth days for Venus to go all the around the Sun. That means that a day on Venus is a little longer than a year on Venus. Since the day and year lengths are similar, one day on Venus is not like a day on planet Earth. Here the Sun rises and sets each day, but on Venus, the sun rises every 117 days Earth because the amount of Sun gravity is centripetal and very important than the centrifugal fundamental thermodynamic force. Although Neptune has the highest fundamental thermodynamic centrifugal force

MOSTINI Planet Next Level

in the range of Jupiter through Neptune. The (size, mass, Entropy) of Neptune, Uranus, Saturn, Jupiter are increasingly huge. Although they have a slightly high amount this range of centrifugal of the fundamental thermodynamic force, the centripetal and gravitational exerted by Jupiter are stronger and leads Neptune and Uranus to perform in the direction of clockwise like Venus, the gravitational centripetal reactivities of our planets. However, the rotation of all the planets in our system are governed by the centrifugal thermodynamic force counterclockwise $F=m\omega^2 r$

The Neptune universal scale reactivities displays highest thermodynamic in comparing the distance from Jupiter to Neptune, lowest mass, and lowest gravitational force From Jupiter to Saturn. And highest thermodynamic force. In the limit of spherical shape all planets of universe, in equilibrium state and the planetary Corona, the amount fundamental thermodynamic centrifugal force equals the amount gravitational G centripetal force. The Entropic force equals the fundamental thermodynamic force.

The planet Mercury is the smallest planet of our

MOSTINI Planet Next Level

Solar system with highest solely and surroundings amount of fundamental centripetal Thermodynamic force. And the highest general (Solar system) centripetal gravitational force. Mercury is the closest planet to the Sun at a distance about 36 million miles (58 million kilometers) or 0.39 ALT. Obviously, one day on Mercury takes 59 Earth days because The General (Solar system) gravity, the centripetal force is the highest and very superior than the fundamental thermodynamic centrifugal force. The centripetal Gravitational G force is higher and very superior> than the centripetal thermodynamic force. One day on Mercury takes 175.97 Earth days.

18 Why All The Planets Of Our Solar System Turn in the Same Direction Except Venus, Uranus, and Neptune

All planets rotate on their axes in an Anti-clockwise direction, but Venus and mercury rotate in clockwise direction. Obviously, Venus rotates clockwise in retrograde once every 243 Earth days - the slowest rotation of any

MOSTINI Planet Next Level

planet. If we consider the universal parameters of the functioning of the universe, we realize that the size, the mass of plants and stars are fixed by the reactivities of the fundamental thermodynamic force.

We know the fundamental thermodynamic force and gravitational force are inversely proportional, the universal celestial entities in equilibrium displays the equilibrium force between the fundamental the centripetal gravitational and Entropic force and the centrifugal fundamental thermodynamic force. Obviously, the fundamental thermodynamic centrifugal force fills the entire universe, so we all have felt it when we are in a car taking the corner at too high speed. It is at the origin of several fairgrounds. It explains the flattening of the Earth, planets, dwarves, Stars on the axes of the poles and circular and round shape of all celestial entities.

$F = m\omega^2 r$

www.ingramcontent.com/pod-product-compliance
Lightning Source LLC
Chambersburg PA
CBHW070618220526
45466CB00001B/43